Using Wireless Communications in Business

Using Wireless Communications in Business

Andrew M. Seybold

Foreword by John W. Seybold

VAN NOSTRAND REINHOLD
An International Thomson Publishing Company

New York • London • Bonn • Boston • Detroit • Madrid • Melbourne • Mexico City
Paris • Singapore • Tokyo • Albany NY • Belmont CA • Cincinnati OH

TRADEMARKS

The words contained in this text which are believed to be trademarked, service marked, or otherwise to hold proprietary rights have been designated as such by use of initial capitalization. No attempt has been made to designate as trademarked or service marked any personal computer words or terms in which proprietary rights might exist. Inclusion, exclusion, or definition of a word or term is not intended to affect, or to express judgment upon, the validity or legal status of any proprietary right which may be claimed for a specific word or term.

Library of Congress Catalog Card Number
ISBN 0-442-01877-0

I(T)P Van Nostrand Reinhold, an International Thomson Publishing Company
ITP logo is a trademark under license.

Printed in the United States of America

Van Nostrand Reinhold
115 Fifth Avenue
New York, NY 10003

International Thomson Publishing
Berkshire House
168-173 High Holborn
London WC1V 7AA
England

Thomas Nelson Australia
102 Dodds Street
South Melbourne, Victoria 3205
Australia

Nelson Canada
1120 Birchmount Road
Scarborough, Ontario
M1K 5G4, Canada

International Thomas Publishing GmbH
Königswinterer Strasse 418
53227 Bonn
Germany

International Thomas Publishing Asia
221 Henderson Road
#05 10 Henderson Building
Singapore 0315

International Thomson Publishing Japan
Hirakawacho Kyowa Building, 3F
2-2-1 Hirakawacho, Chiyoda-ku
Tokyo 102
Japan

ARCFF 16 15 14 13 12 11 10 9 8 7 6 5 4 3 2 1

Library of Congress Cataloging in Publication Data

Seybold, Andrew M.
Using wireless communications in business / Andrew M. Seybold.
p. cm.
Includes bibliographical references and index.
ISBN 0-442-01877-0
1. Business—Communication systems. 2. Computer networks.
3. Wireless communication systems. I. Title
HF5548.2.S4373 1994
651.8—dc20

94-7271
CIP

Contents

Foreword

The story of the evolution of society is the story of the evolution of communication—the development of languages (from gestures, signals and pictures, to the spoken word) passed on from one generation and tribe to another. Eventually the ability to record observations and insights was developed and, still later, the ability to transmit those recordings not only over time, but through space as well.

Pictographs became alphabets, capable of defining not only things, but ideas and concepts. Music and art evolved, too, permitting the communication of moods, whims, fancies, and sentiments enhancing the ways of expressing subconscious feelings.

Mankind was given five, or perhaps six, senses to work with: touch, sight, hearing, taste, smelling, and intuition. All of these led to understanding. We learned to describe what we saw and to record it—not as sounds, but as words and pictures. Much later we learned to record what we heard, and to transmit it over great distances as well as to store it and make it available for future access. Most recently, we have learned not only how to record what we can see and hear, but also to store it and transmit it over great distances.

With the proliferation of such information, two significant problems have been presented. One is how to locate quickly desired or relevant portions of stored information. The other is how to minimize the space it takes to store information, and the length of time it takes to transmit that information. One of the most fascinating aspects of communication technology, and that which relates specifically to the use of computers, is data compression, both for transmission and for storage of information. The other area in which computers are useful is to help expedite the access of information.

Before moving on to these most sophisticated areas, let us backtrack a little to describe the ways in which mankind has learned, over the centuries, to store and transmit the written word.

Mankind moved from pictographs on cave walls to scratches on parchment or papyrus, to alphabets, to engraving, to movable type, to automated typesetting machines; and moved from smoke signals to drum beats, to hand-delivered written messages, and then, with the advent of electricity, to the telegraph, and the Morse code.

The Morse code is still with us. It was developed in the 1840s by Samuel F. B. Morse and used initially for telegraphy and then adapted for wireless communications. For better or worse, Morse coding conventions, as adapted by Baudot for teletype (32 combinations), and modified by Walter Morey (64 combinations) for teletype, account for six of the "bits" in 128 of our 256-character EBCDIC or USASCII eight-bit "bytes" common to all computer transactions.

When, in the 1960s and 1970s, photographic typesetting machines sought to create characters by exposing blips of light on a screen, run-length encoding techniques were developed to compress the amount of data the computer program was required to handle. And much more sophisticated approaches were taken in the 1980s to describe the shapes of the outlines of individual type characters.

Similar techniques were developed to reduce the amount of data needed to describe the contents of a scanned image (character, drawing, or photograph). Some techniques are simple, as in the case of the repetition of all-black or all-white areas. Others are quite sophisticated, dealing, as they do, with the amount of "information" the eye needs to see in a given application in order to recognize a shape, pattern, or color.

Scanned input involved a transition from the receiving of analog information (more or less, higher and lower) to digital (yes or no, 0 or 1) and, of course, reproducing its output as well, whether by representation on a TV-type monitor or by means of a dot matrix or laser printer.

In the meantime, the representation of sound was addressed. Perhaps the beginning of modern sound technology could be traced to the development of the motion picture. In the early 1900s, we learned that by drawing pictures on a pad and flipping through them. The idea of motion could be imparted to still pictures if successive versions were presented, letting the eye reconcile their minute differences. Similarly, we learned that sound could be cut into tiny slices and recorded digitally, letting the ear and the mind reconstruct the original version.

In like manner, there are similar opportunities for data enhancement or modification, working with computer-stored "bytes" or "words" rather than with the original input. Now it is no longer possible to assume that the picture seen on the tube, on the screen, or in print, is an actual photograph, or that the sound heard was ever really played or uttered. Techniques for certifying or authenticating the fidelity of that which is seen or heard will need to be developed when it has been so reproduced and stored.

Despite the evolution of these civilizing techniques to enhance our ability to communicate and to understand the world of facts and ideas all around us, we have only begun to apply evolving computer technology to the field of communications. Often, perhaps most often, invention has been born out of necessity. Now it may be the case that imagination may need to be directed toward the effective and productive use of capabilities that are now there (or almost there) for the taking.

Moreover, there now exists the ability to simulate—to construct and to play out hypothetical situations—in order to determine whether the likely consequences of a given course of action would be beneficial or destructive.

For most of us, the task is to understand what technology is already available in order to discover or conceive of how it may best be used. To accomplish this, we will need to look at these new technological possibilities with great care. We need to take the time to understand these building blocks—how they work, how they might fit together—how they can be made to relate to one another, and how they can be adapted or developed to improve our ability to communicate in the highest and most efficient ways.

Too often we get carried away with technical possibilities, but we need to understand the technology in order to adapt it to our needs. Never before have we had available such an embarrassment of technological riches in the field of communications. To make the most of this technology, one must first understand it—how it works, what it portends. This is an assignment not just for the specialist, but for all of us!

Technological improvements or developments often set into motion forces and influences with unforeseen ramifications. The public as a whole can benefit or suffer from these new discoveries and applications; hopefully, what transpires will be beneficial and enriching. At the same time, we must not forget the legend of Pandora's box, or of the prophetic words of Mark Anthony—"Now let it work. Mischief, thou art afoot. Take thou what course thou wilt!"

This book is written with the hope that we will indeed be able to plan our future in such a way that all of our lives may be enriched. In order to do that, however, we shall need a broader understanding of the available options and opportunities.

John W. Seybold

THE SEYBOLD FAMILY

John Seybold, the father of the three Seybold Siblings (Jonathan, Andrew, and Patricia) was the first to form a company dedicated to computerizing the typesetting industry (ROCAPPI, Research on Computer Applications for the Printing and Publishing Industries).

In 1971, John Seybold sold ROCAPPI and formed a newsletter and consulting business with Jonathan to advise others in the industry. The original *Seybold Report* is widely known as the definitive publication for this industry. In 1976, Patricia joined the family business and *The Seybold Report on Office Systems* was launched. In 1983, Andrew joined the organization and the *Seybold Report on Professional Computing* was initiated.

When John Seybold retired, Patricia and Andrew left the parent company; Patricia formed Patricia Seybold's Office Computing Group which published the Office Systems publication, and Andrew established The Seybold Group which published the Professional Computing newsletter.

As John was instrumental in helping to establish computerized typesetting, Jonathan was deeply involved in forming what has become the desktop publishing industry. Patricia is credited with formative work in the area of office automation. Andrew is acknowledged by the industry as a driving force in the portable computer industry and is part of the advance guard of the wireless revolution.

Acknowledgments

No book is the work of a single writer. Many others contribute to the effort before a manuscript is a finished product; many have contributed to these pages both directly and indirectly.

Many people in the computer and communications industries who are helping to shape the wireless industry have provided valuable information, insight, and encouragement. They are too many to mention by name, but I would like to thank each of them for their contributions to this work.

I would also like to express my appreciation to my father, John, who has been behind the scenes during my career as a writer, encouraging, teaching, and helping me through the rough spots, and to Linda, my editor, partner, and wife, without whom this book and our newsletters would not be possible. Her work, along with that of the staff of VNR has transformed my thoughts and ideas into a solid, readable book which we all hope will impart readers with an understanding of the capabilities and promise of wireless communications.

1

Introduction

Every field of endeavor tends to develop its own specific language and "techno-jumble." The computer industry's jargon is quite foreign to the non-computer literate. The meanings of its terms and phrases are not self-evident, and they can be confusing. Consider, for example, words like RAM, ROM, BIOS, boot track, VGA, Super VGA, MCA, ISA, EISA. Even so, people in the industry tend to inflict its jargon upon the uninitiated, thereby making the newcomer's first forays into the world of computers all the more difficult and frustrating.

The communications industry has its own language with terms that are foreign to most people, including many who are involved in the computer industry. Words such as simplex, half-duplex, duplex, repeater, sub-audible tone (PL), sensitivity, dB gain, VSWR, mod, deviation, spread-spectrum, ASCB, baseband, mux, CDMA, TDMA, and watts are all bantered about by the wireless segment of the communications industry. The folks who deal with the "telecommunications side of the house" use terms unique to their specialty, such as 2-wire, 4-wire, digital switch, PBX, mux, to list a few.

Understanding that the nature of the communications industry is both like and unlike that of the computer industry is important. In its "like form" the communications industry is comprised of companies that design, build, and market equipment directly to the end user and business community. Companies such as Motorola, Ericsson, and NEC may be compared to the likes of IBM, Digital Equipment Company, and Apple.

Each of these companies (as well as many others) is wrestling with the same issues as the computer companies—technology advances, products, cost of materials, cost of sales and marketing, and customer allegiance. These companies are international in scope, selling equipment into many different regions, and each region has its own rules and regulations. Each company tries to provide the best possible products and services at the best possible prices.

When comparing Motorola with IBM, one finds rather striking similarities—each is the largest market-share holder in its respective field and each was initially striving to sell directly to the customer. Each demanded, and got, a premium for its products. And each has recently undergone changes in product offerings, ways of marketing, and in its organizational structure.

However, there is an aspect of the communications industry that is not yet found in the computer industry. In addition to having to deal with manufacturers and customers, there is a third, extremely important segment of the communications industry known as "service providers." While some service providers are also equipment manufacturers (Motorola) and telephone systems providers (AT&T and the Baby Bells), some service providers are organizations that do nothing but provide back-bone communications equipment, obtain a frequency or set of frequencies from the Federal Communications Commission (FCC), and sell time on these systems. They are not—or have not been—in the business of manufacturing or developing hardware.

A parallel can be drawn between these providers of radio channels and telephone companies. Radio carriers and telephone companies both provide paths and systems that route information to the proper destination. At present, telephone companies accomplish this with wire (a resource that can be increased as needed, but at some cost), whereas radio service providers deal with a scarce resource, the radio spectrum.

Both carrier groups (phone companies and radio carriers) are regulated by the federal government and by the states in which they do business, but radio carriers are resource-constrained while telephone carriers are not.

Understanding how this third segment of the industry, the transport layer providers, works, is regulated, and how it stands to gain (or lose) from the proposals now before the FCC is important. These companies can be the computer industry's greatest allies within the communications industry; they can also be its biggest adversaries. The computer industry must make the choice.

A SHORTAGE

Currently, a severe shortage of radio spectrum exists, despite all the advances in technology. The federal government is the biggest user of spectrum in the United States, and the FCC has no jurisdiction over these frequencies. The amount of spectrum available to the FCC for allocation to other users is what is left after the government gets what it wants. (See Chapter 2 for a discussion of the spectrum.)

Meanwhile, the FCC is in the process of re-allocating up to 220 MHz of spectrum in the 1.8 to 2.5 GHz range for "Emerging Technologies" or "Personal Communications Services" (PCS). The FCC allocates frequencies by passing a new rule designating the channels for use, and defining who may use those channels. A proposal for rule-making is the first step in a lengthy process. Once a proposal has been released, comments are solicited from interested parties, comments on the comments are then entertained, debate is undertaken, and, sometimes, a new rule is passed.

In the case of the "emerging technologies" proposal for rules, there has been one ruling defining the band to be used, and many more proposals for specific uses of the frequencies within that band are yet to come.

A battle between existing users of these channels (police, fire, utility companies, and others) and the computer and communications industries is imminent, and this has all the makings of a major political battle. Historically, the FCC has encouraged organizations with common interests and goals to form industry associations that, hopefully, represent the majority view of that specific industry.

Some of the organizations formed to represent their industries are: *Telocator*, serving paging, cellular, and SMR providers; The Associated Public Safety Communications Organization (*APCO*), serving police and public safety sectors; The Forestry Conservation Communications Association (*FCCA*); The International Association of Fire Chiefs (*IAFC*); The Special Industrial Radio Services Association (*SIRSA*); The National Association for Business and Educational Radio (*NABER*); and the Cellular Telephone Industry Association (*CTIA*).

Most of these organizations not only serve as representatives of their member organizations, but, with FCC sanctioning, they have also undertaken most of the frequency coordination activities for their respective industries. Frequency coordination is the task of finding a usable frequency within the allocated band for a new applicant for radio service.

The frequency coordination process works something like this: RA Company decides to purchase a voice radio system for use by its field service department. It calls a two-way radio vendor who examines its requirements and chooses the proper radio band and type of system that will best meet the coverage needs of the company.

The next step is to complete a series of forms identifying the company, the location of the base station, service area within which the base station needs to be able to communicate, the type of equipment proposed, power and antenna configurations recommended by the vendor, and other pertinent details. One of the forms is a formal license application to the FCC. Once completed, these forms are sent or taken to the frequency coordinator of NABER, the industry organization for the business radio service.

If a specific frequency has been requested, the coordination committee looks at that channel, determines who else is using it, and determines if there is potential for interference. If the coordination committee finds that the frequency requested is the best one available (this is rarely the case), it signs the coordination papers and returns them to the business.

If a frequency is not specified by either the business or the vendor, the committee will refer to its database and recommend what it believes to be the best frequency for the company. In reality, with the present shortage of frequencies available within each service, the business radio committee will most likely recommend using a "shared radio system" of some type.

This process has been effective for a number of years and the FCC honors the requests of these organizations. On the other hand, if a company files an application for a frequency in one of these bands and does not go though the coordination process with the proper industry organization, the application will be returned by the FCC without action.

This explanation of frequency coordination is relevant to the computer industry for two reasons:

1. Because industry organizations have voluntarily taken the coordination burden off the FCC, the FCC tends to listen to them.
2. Since these organizations represent the members of a given radio use community, they represent the collected clout of that industry.

In addition to speaking for their own industries, the organizations cooperate (at least in theory) with each other for the betterment of all. Each employs attorneys and lobbyists who know their way around within the FCC structure and who follow the FCC's every move. Because the communications industry is so dependent upon a very scarce resource (radio spectrum), it has put in place a number of different and effective ways of tracking activities within the FCC, from both an official and exploratory stance.

These organizations are well established, highly organized, highly funded, and very protective of their own "turf." The computer industry is about to claim a large amount of this very limited and valuable resource. It would be foolish to expect those who now preside over this spectrum to roll over and give up a significant share simply because of some "wild" ideas about wireless communications and personal communications services.

If computer manufacturers and users are to work with the communications industry, they must learn to understand and use the language of communications in order to explain their terms, and their needs, to the communications community. Inability to understand each other will result in many aborted attempts at the marriage of these technologies—two technologies that are acknowledged to be the cornerstone of the mobile computing era, and, beyond that, of the entire communications world of the future.

We are on the verge of a new technological revolution—the merging of the communications, computer, and information industries. The result will be a new way of providing communications infrastructure, information, and capabilities to anyone, anywhere. The technology necessary to provide the backbone and delivery devices for this new and exciting communications structure is either available today or will be forthcoming in the very near future.

It is not possible to pick up a newspaper or watch the news on television without being reminded that this communications revolution is coming to a device near you "Real Soon Now" (RSN). The truth of the matter is that the revolution is taking place faster than most people realize. It will be upon us before we are prepared for it and before we have been able to digest the premises behind it.

We will soon have to make choices about who will supply our dial tone in our homes. Will it be our traditional telephone company? Our cable TV company? Or AT&T combined with McCaw Cellular to bring dial-tone not only into our house, but anywhere we go? Will our video choices—which have already exploded from three or so channels of network programming to well over 100 channels brought to us by our local cable company—now expand to 500 or more? What will these channels be used for? Who will provide the service, and what will it cost? And most importantly, do we really *want* and *need* these services?

Delving into the questions and technologies that will affect our personal lives is beyond the scope of this book. However, it is important for the reader to realize that just as communications and computing are merging and evolving in the workplace, they are rapidly coming together in the home. Beyond these technological possibilities are the changes that can blur the distinction between home and office, while "time on the job" becomes less important.

When all of these voice and data communications tools are in place, we will be able to access our information anywhere, anytime, and communicate with whomever we need to communicate. This may render things such as formal offices with walls and plastic plants and reproduced artwork unnecessary. Instead, offices can be anywhere, and our days can be a mixture of business and leisure time: we will be able to take our children to the zoo and at the same time stay in touch with our work world—if that is what we want to do. The merger of communications and computing is much more than the marriage of technologies. It is about access and availability, and we will soon have the technological means of experiencing the opportunities and dilemmas associated with the question of access and availability, if that is what we choose.

THE FUTURE

Many people can paint a picture of the future. Trade shows and conventions feature video tapes depicting the future. In each video, the lead character is distant from his or her place of work, and he or she is intimately involved in solving a problem by making use of a communications device of the future.

Such visions are fascinating. Most people can identify with being able to work wherever they are, however they are occupied, and fully expect this to happen over time. But, technology notwithstanding, many obstacles must be overcome before these visionary capabilities are achieved.

Wireless communications, the chief enabler of mobility, is limited by both physics and regulations. Depending on the form of the information people need to be able to communicate (voice, data, video, etc.), bandwidth will be required for implementation. Currently, not enough bandwidth is available to permit the realization of visions depicted in videos, nor is enough bandwidth available to enable all to communicate with whom they want, when they want.

USING WIRELESS COMMUNICATIONS IN BUSINESS

This book is not about communications as depicted in futuristic videos. It is about today. It describes what is possible today, what the limitations are, and how to make use of today's technology to solve communications and computing-intensive problems without waiting for the future. This book is also about issues such as for whom these new technologies are best suited, when they should be implemented, and when one should wait for the next generation of products and services.

The following pages explore the various technologies that comprise the products and services that are available and offer a discussion of some of the historical events and mentalities that will either help or hinder the implementation of the technologies.

Those who have already embraced the various technologies and are making use of them on a daily basis now wonder how they worked without them only a few short years ago. While the use of these new tools makes some people more productive, this technology may not be necessary for everyone. In fact, some of this technology may be rejected by the general public until it becomes clearer that there really is an advantage to using it.

It is just as important to understand the limitations of what is possible as it is to understand the possibilities. Those who are ready to experiment with or fully embrace the use of these technologies will do so more quickly, and with more relish, if they fully understand both the benefits and the limitations of what is available. Unfortunately, human nature often causes those who provide the tools to represent their use as being like child's play and, for some providers, using these tools may indeed seem easy. But for many, using certain highly-touted applications may prove to be more trouble than they are worth.

The intent of this book is to impart enough information and knowledge to help the reader understand what is happening in this convergence of technologies and then to be able to make decisions regarding the use of these technologies. The decision to become wireless-enabled should be made with an understanding of what is possible and what is not.

This is not a "how to" book; rather, it is a "what if I want to?" book. It is designed to educate and impart an understanding. Such an understanding should enable us to discern what is real as well as to develop a healthy skepticism of the hype and over-simplification that prevails when scientists turn technology over to marketing and sales organizations.

USING THIS TEXT

In this text, the reader will encounter general explanations of a number of technologies that may be unfamiliar. A technical section is provided in Appendix B that explains in more detail specific technologies employed in wireless communications systems. Readers should refer to this section whenever a mentioned technology piques their interest, or if they are interested in learning more about a specific technology.

Efforts have been made to offer general explanations of terms within the text. More detailed definitions can be found in the Glossary.

The following three chapters present an overview that is designed to introduce the uninitiated to the world of wireless communications.

Chapters 5 and 6 delve into the challenges facing the computer and communications industries as those industries strive toward the goal of "anytime, anywhere" computing mobility. Meeting these challenges will result in a revolution in the way we do business.

Understanding these challenges will help individuals and businesses to make decisions about when and what to implement for a business advantage.

Chapter 7 provides detailed information about each option available today or in the near future. With the knowledge gained from the preceding chapters, the reader will be able to assess the merits of each system to determine whether it promises to provide a solution to his or her business needs.

To assist in the evaluation of the systems as they apply to your needs, Chapter 8 supplies a recap, and Chapter 9 offers guidelines to be used in your selection and implementation process.

2

Frequency Allocations and Uses

This chapter summarizes how the radio spectrum has been allocated over the years as new technologies have been developed and new uses have been identified. Within the United States, the communications industries must either live within the limitations of the established structure, or request new allocations, or re-allocations, from the FCC. Figure 2-1 depicts the radio spectrum and how it is allocated.

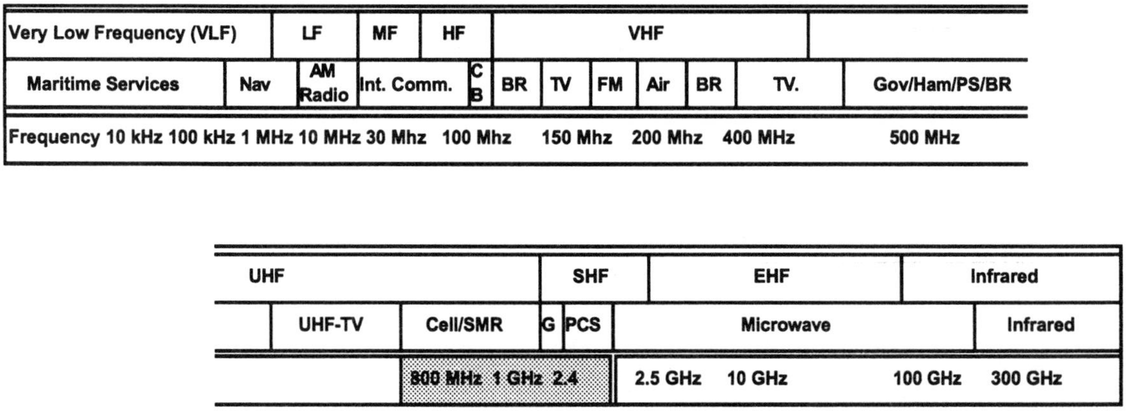

Very Low Frequency (VLF)		LF	MF	HF		VHF						
Maritime Services	Nav	AM Radio	Int. Comm.	CB	BR	TV	FM	Air	BR	TV.	Gov/Ham/PS/BR
Frequency 10 kHz 100 kHz 1 MHz 10 MHz 30 Mhz 100 Mhz 150 Mhz 200 Mhz 400 MHz 500 MHz											

UHF				SHF	EHF	Infrared
	UHF-TV	Cell/SMR	G	PCS	Microwave	Infrared
		800 MHz 1 GHz 2.4			2.5 GHz 10 GHz 100 GHz	300 GHz

Figure 2-1. *The Radio Spectrum*

Radio spectrum that can be used with today's technologies is *fully allocated.* New allocations will require new technology that is capable of using frequencies that have been considered unusable or clearly beyond anticipated capabilities.

Re-allocation is difficult to accomplish. If building equipment that would work over the entire spectrum were possible, reassigning frequencies in a logical sequence with groupings of like services would be a simple matter. Unfortunately, with today's technology, radio transmitters, receivers, and antennas are built to function in a specific range. Moreover, the costs associated with re-allocation, and the resulting need to replace equipment, appear to be prohibitive.

To understand the allocation structure and its implications for providers and users, one must understand how the allocation structure developed.

COMMUNICATIONS CARRIERS AND SYSTEMS

Many types of communications links are available to users. *Wired links* include telecommunications services over traditional landline circuits and fiber-optics. Wireless systems include radio frequencies (RF) in Ultra High Frequency (UHF), Very High Frequency (VHF), Low Band (LB), and microwave bands. Some systems even make use of satellite links. These links make up the "transport" layer over which voice, data, and video are sent.

To understand the relationships and precepts of industries that rely on carriers, it is helpful to know about the radio frequency spectrum—who may use it, and who controls it. This spectrum is a finite resource, like oil. When it is finally gone (all allocated) there is no more available. However, advances in technology provide us with additional usable spectrum just as steam oil drilling provides oil from a dry well. Oil fields can be depleted forever and radio frequencies cannot, however there may be substitutes for oil, but not for radio spectrum. Both these industries, however, will soon run out of their respective resources even with technological advances.

Some of the spectrum (such as frequencies used for AM and FM radio and broadcast television service) provides access to *programming* (data) of general public interest. Some is allocated for communications among *super-secret military and government organizations.* Some is used by *public safety agencies* for police and fire department communications, some by *business and industry*, some by *amateur radio operators*, and some for *satellite communications* (voice, data, and video).

There are also frequencies for *international communications*, *international broadcast stations*, *off-shore oil rigs*, *pipeline control*, *cordless telephones*, and even *garage door openers!* In each case, the need was discovered (or invented) and, within the United States, the FCC determined exactly how much spectrum in what frequency band was required for the specific use. After the proper legal process, the allocations were made.

The *AM broadcast band*, *FM broadcast band*, and *TV bands* each consist of a number of different channels or frequencies on which organizations are licensed to conduct one-way radio and television broadcasting.

The *Very High Frequency (VHF) two-way radio band* consists of a group of frequencies (150–174 MHz) authorized by the FCC to be used by various types of organizations needing paging and/or two-way radio capabilities. Just as you will find a rock-and-roll station next to an "elevator music" station, next to an all-news station within the FM broadcast band, so, too, will you find police, fire, marine, paging, taxi, business, and other types of radio services within the VHF band of frequencies or channels.

FREQUENCY PROPAGATION

At first, only a small amount of radio spectrum was available for use. The technologies of the day could only provide transmitters and receivers that worked at very low frequencies (a radio frequency is measured in meters and in Hertz). In the "old days," allocations were made in the 80-meter band, or at 3.500–4.000 MHz.

Different frequencies have different characteristics. As a result, some frequencies are better suited for long distance communications; some are better for a shorter range, or for in-building coverage. The general rule is that the higher the frequency in MHz, the shorter the range. At the lower frequencies, the point-to-point range of a radio transmission is much further than it is at a higher frequency. (Compare, for example, 150 MHz versus the 800-MHz cellular mobile telephone band.)

A close look at the radio spectrum allocations above 30 MHz (10 meters) reveals some interesting things. The FCC, years ago, organized frequencies by type of service. Therefore, all FM broadcast stations are located within the 88–108-MHz spectrum. This practice of grouping started in the 30–50-MHz band (called the "low" band by communicators). Within the 20 MHz of this band, the FCC allocated a portion for fire service, a portion for police, groups of frequencies for non-public safety business, and other frequency groups for other types of communications.

As technology advanced and demand for frequencies increased, the FCC followed the pattern established in the low band when it allocated the next set of frequencies. The Very High Frequency (VHF) band extends from 150 to174 MHz and is assigned to many different groups of radio users including police, fire, taxi, business, and marine two-way radio systems, as well as one-way paging systems. (Two-way radio is defined as the ability to transmit *and* to receive radio signals; however, a one-way radio either transmits *or* receives, as in AM and FM broadcast systems and radio paging applications.)

Between the time the 30–50-MHz band and the VHF two-way radio band was allocated for two-way radio, both FM broadcast and television stations came into being and were assigned frequencies above the low band, and some below and directly above the VHF band.

Time, technology, and demand kept pace for a few years and the FCC opened more frequencies in higher and higher ranges. TV channels above channel 6 were given the 174 to 216 MHz range, and the government usurped allocations in the 200–400-MHz range. The land mobile service demand for frequencies outstripped those available in the VHF band, and they were assigned the 450–470-MHz band. This band was divided into sections, as was the 30–50-MHz band. Each was assigned its own little block of spectrum between 450 and 470 MHz—police, fire, taxi, paging, trucking, point-to-point control, and even remote audio feeds for the broadcast industry.

This method of frequency assignment has had serious implications for fire and police agencies, among others. Because of increased demand for channels, allocations for these agencies were made from different ranges of the radio spectrum. In many cases, it became a challenge for police cars or fire trucks from two different towns to talk to each other.

Over the years, as more and more radio frequencies have become available, there has been little, if any, effort to re-allocate channels within the same bands. The reasons have to do with the economics of relocation rather than any omission on the part of the FCC. The cost to the user to re-locate to a different radio band or channel can be very high. To move to another band is not merely a matter of tuning to a different radio station. Just as an AM station cannot be received on an FM radio and vice versa, a radio in the VHF band cannot receive or transmit in the UHF band. Instead, the two-way radio user must replace entire transmitter and receiver units, install new antenna feedlines, change antennas, and sometimes the towers that support the antennas must be replaced.

Several times in the past twenty years, the FCC has looked at frequency reallocation. Each time, the user communities have protested the plans because of the high cost of equipment change-outs.

Above the 450–470-MHz band, the FCC assigned Ultra High Frequency (UHF) TV channels (13–72). Because the frequencies are much higher than those used by the VHF TV stations, the distance that a signal can travel is significantly less for the same amount of power. Therefore, UHF stations are generally more local in nature and none of the major networks have been concerned with these channels.

Interestingly, while the FCC was debating how to allocate the next generation of technically-accessible spectrum (the 800–900-MHz frequencies), there was a critical need for additional two-way radio channels. The FCC embarked on a historic first by permitting two-way radio operators to make use of UHF TV channels that are not in use within their metropolitan areas.

The idea was simple. In Philadelphia for example, a UHF TV station transmits on channel 19. Channels 18 and 20 are not in use because adjoining channels are kept clear to give a buffer, or protection zone, against interference. Each television station uses 6 MHz of spectrum for its video and audio carriers, and two-way mobile uses only about 15 KC of spectrum (including audio and protection for the adjacent channels). A total of 66 two-way radio channels could be assigned for each television channel that was not in use in the given area.

The FCC began to license two-way radio systems in the UHF-T band (previously called the UHF-TV band) in the 1970s, but it imposed restrictions in power and area coverage so that interference between two-way radio stations and television stations would be kept to a minimum. For local area coverage by users of two-way radio, the decision to share television channels with mobile radio has worked well—

but it doesn't even begin to handle all the requests for frequencies that are pending before the FCC.

THE GROWTH INTO 800 MHZ

The next group of channels allocated for two-way radio, one-way paging, and cellular telephone systems was spectrum in the 800–900-MHz area. A few short years before, these frequencies were considered "microwave," or point-to-point channels only. Today, they are widely used for telephone, paging, police, fire, local government, business, and other communications services.

In an effort to better utilize this portion of the spectrum, the FCC granted frequency allocations in the 800–900-MHz band in a different manner. Some services, such as police and fire departments, received the same type of "block" spectrum allocations as they had in the lower bands. The difference is that in addition to cellular phones, the FCC authorized a new service that is provided by the Specialized Mobile Radio Service (SMR).

SMRS

The technology used for this new service is a cross between that of standard two-way radio and cellular systems. Each system is granted a license for a number of channels ranging from a low of 5 to an initial high of 20 (depending on the number of mobile units that were to be used in a system—the more mobile units, the more channels). Each provider of service was then free to sell equipment and/or service on these channels for use by customers who need a "closed" two-way radio system, such as a fleet operations or service companies. Telephone interconnect is also permitted, giving the users a combination of two-way and telephone service within their own geographic area.

The theory of SMR operation is that one of the channels is designated as the "calling" channel and all mobile and portable radios "listen" to this channel (using digital signaling techniques . . . the user hears nothing). When a call is made, the data stream assigns one of the vacant channels to the customer and the radios in each vehicle are electronically "tuned" to the vacant channel so a "private" conversation can be held. Once the radio contact is over, all vehicle radios return to the main channel and wait for the next call.

SMR service has developed far beyond the simple description provided here. True telephone interconnect and expanding service areas have made SMR a competitor with cellular phone systems in some areas. Several SMR organizations have recently come together to provide a nationwide SMR system that will be capable of data communications as well as voice. This will put them in competition with companies that want to use cellular telephone systems to transmit data, as well as companies such as ARDIS and RAM Mobile Data that have established nationwide, data-only radio networks.

3

Wireless Local Area Networks

This chapter looks at what can be implemented today in the world of wireless data communications. It is a snapshot in time as we move forward toward our totally wireless world. Futures may be more fun to contemplate, but providing today's solutions requires the use of today's technologies.

SETTING THE STAGE

When one thinks about "wireless computing," what generally comes to mind is an ultimate vision of being able to access both voice and data communications circuits no matter where the user is, and no matter what the user is doing. This vision has been expressed by people representing a diverse group of companies, but the reality is that this type of all-things-to-all-people computing is still a number of years away.

Many diverse groups must come together to make this vision of wireless computing a reality. The federal government (in the form of the FCC and other regulatory agencies), communications service providers, communications equipment suppliers, computer hardware and software suppliers, and information service providers must all share the same vision and learn to work together toward the implementation of that vision.

In the process of moving forward toward this utopian goal, the most-asked questions are about when to start planning for wireless computing and when to start implementing a strategy. The following summary delineates what is available today and provides guidelines for implementing systems that will work now and will continue to work well into the future.

WIRELESS PROVIDERS

Currently, implementation of both local area wireless and wide area wireless systems—either as data-only or with mixed data and voice is possible. At present, the local area connections are all designed to be extensions to existing Local Area Networks (LANs), but the Wide Area Networks (WANs) can be used to send and receive voice and/or data from virtually anywhere within the United States and beyond.

As action is taken by the FCC on the unlicensed portion of the Emerging Technologies (PCS) bands (1910–1930 MHz), other uses will be developed for in-building and on-campus wireless communications. For the moment, the companies offering products in this specific market have targeted LAN extensions.

LOCAL AREA SYSTEMS

Most of the products on the market today are designed to be used for extending existing LANs and, therefore, "look" exactly like a hardwired Ethernet or TokenRing node to the network and the server. The difference is that they are using infrared and Radio Frequency (RF) technologies.

The bulk of such wireless systems is being installed in place of direct-cabled networks for one of two reasons:

1. The location of the desired node makes running wire from the node to the rest of the network expensive or technically impossible;
2. The purpose of the wireless portion of the LAN is to add coverage to common areas where groups will be gathering and where LAN connectivity is important (conference rooms, for example).

Each of these uses of wireless LANs requires a different approach to the installation and the equipment used as the "node." (For purposes of this discussion, a node is a computing device that is attached to the network to make use of network services such as e-mail, printer access, and file and data access.)

If the purpose of the wireless link is point-to-point network expansion, the system will be used with standard desktop PC hardware and will, most likely, include a network add-in card that will install much as wired network cards do now. Even the software used to control these wireless nodes looks and feels like typical network client software and is installed as though an additional wired node is being added.

This type of wireless LAN extension is the easiest to install and maintain since the network "sees" the wireless node exactly as though it is part of the wired system.

FREQUENCIES AND TECHNOLOGIES

Several different types of infrared and RF systems can be used for a wireless LAN extension. Figure 3-3 lists the technologies and companies involved in providing equipment for this type of point-to-point network extension.

CHOOSING A SYSTEM

Wireless LAN Guideline Number 1: Determine the physical location of the system. Choice of system will vary based on the following criteria:

- What is the distance between the wireless node and the closest network point?
- Is the node line-of-sight to a network connecting point?
- In what type of structure will the system be located? Steel and concrete, metal, plaster, other?
- What type of data needs to be exchanged between the node and the network?

INFRARED

If the distance between the node and the rest of the network is less than 800 feet and is line-of-sight, an infrared system such as that sold by Photonics may be the best solution. The advantage of infrared over RF is that infrared will not penetrate the building—it will, therefore, be a more secure link. However, the speed capabilities at present are limited to 242 KBps (kilobytes per second) which may not be satisfactory if a large amount of data and graphics needs to be moved to this node.

RF CHOICES

There are only a few companies offering RF wireless links that can be used for a point-to-point type of LAN expansion. Motorola offers one that is licensed and, therefore, coordinated with others in the same area, and NCR, O'Neill, and Proxim all offer unlicensed RF solutions.

The Proxim, O'Neill, and NCR solutions use spread spectrum technology in the unlicensed frequency band of 902–928 MHz. There is some concern about the long-term viability of the 902–928-MHz unlicensed band because in addition to unlicensed use, it is also available to four different groups of licensed users. What this means is that unlicensed users must accept any and all interference generated by any of the licensed users and must *not* interfere with them.

For the immediate future, the interference issues should not be a problem in most areas of the country. But those deploying systems in this band need to be aware of the potential for interference and that they—as unlicensed users of this band—have no recourse if they do experience interference.

Meanwhile, there are several other bands that can also be used for low-power unlicensed LAN operations. Two of these are in the 2.4-GHz and 5.4-GHz Industrial, Scientific, and Medical bands (ISM). The FCC has also allocated some of the Personal Communications Service (PCS) spectrum for such use.

In the PCS ruling, the FCC allocated 20 MHz of spectrum for unlicensed voice use and 20 MHz for unlicensed data communications. At the end of 1993, several companies—including Apple Computer—filed motions with the FCC for reconsideration of these allocations. One may safely assume that the bulk of future RF wireless in-building LANs will make use of the new spectrum allocated in the PCS frequency plan.

SPREAD SPECTRUM

The equipment presently available in the 902–928-MHz band uses spread spectrum technologies. These are not as prone to interference from other radio sources as are standard radio frequency technologies. However, as with any radio system that operates in an unlicensed portion of the spectrum, the FCC provides no guarantees from interference. In fact, the band is available to many different types of users including the next generation of digital portable phones, amateur radio systems, and many other low-powered devices. As a stipulation of choosing unlicensed systems, users must acknowledge the possibility that their systems may experience interference problems from other systems, or that their systems may cause interference problems for others (and they will be responsible for correcting those problems).

The idea behind spread-spectrum technology is that it is less susceptible to radio interference than are other radio communications systems. Still, interference can exist even in this environment. Therefore, care should be taken when implementing RF LAN devices in a high radio frequency noise environment. One problem with radio that those without a radio communications background may not realize is that radio interference does not have to be generated by radio equipment in the same band or even in the same area. Radio frequencies can "mix" with others and can be rectified and re-transmitted by some very interesting objects—such as rusty bolts on radio towers, rust in chain link fences, and other metallic surfaces.

The art of determining the cause of radio interference is the subject of volumes of books and scores of articles. This an inexact science—one that has caused much frustration for many well-qualified radio engineers. An important point to remember is that the more radio transmitters in a given area, the more likely it is that a radio interference problem will occur at one time or another.

None of this should dissuade anyone from implementing an RF wireless LAN if it fits into his or her overall plans; and one must be careful about the placement of the LAN, and its relationship to surrounding environment. Most of these systems are "smart" enough to detect channel activity and move to frequencies that are not in use in the given area—all without an operator's intervention.

TWO-YEAR PLANNING

Wireless LAN Guideline Number 2: A wireless LAN connection should not be implemented unless a payback period that is less than two years can be calculated. It is likely that within the next two years the technology for wireless LANs will have changed dramatically. A system located within a large urban area may begin experiencing delays due to radio interference as more and more wireless communications equipment is installed in the area.

CONFERENCE LAN CONNECTIONS

A second type of wireless LAN connection is used in places such as conference rooms where group meetings take place. The reason for placing LAN access in these areas is to permit those with portable computers to use the network during meetings.

This type of connection is particularly well-suited for meetings where data that may be on the network has not been brought into the meeting. In this scenario, any user with a properly equipped portable computer in the conference room could access his or her own data and e-mail messages.

NODES

This type of connection is somewhat more difficult to achieve since conference room coverage is designed not for a specific node, but for any number of different nodes or users. Some of these nodes may normally be connected to other servers in other buildings, or even on other networks that require bridges and routers for access.

This is one area that Novell Netware and several other network operating systems are not yet able to handle. Novell is designed to be a wired network, and therefore it is not equipped to handle "roaming" nodes that appear and disappear at different locations and on different servers.

Proxim has been able to overcome this one-connection, many-node network limitation. Proxim has discovered a way to make Netware "think" that a node connected to a server other than the one to which it has been assigned is really connected to the assigned server.

All the other providers of wireless network services, including the Photonics infrared systems, require a user to log into his or her own server. If the wireless node connects to a different server within the network, the network may get "lost" trying to find the path between the wireless node and the user's own server.

If a building is wired with a single server, and if the meeting rooms are wired with a LAN connection—either infrared or RF—no problems should occur. However, if more than one server is on-line in a given network, there could be a connection problem.

CONFERENCE LAN HARDWARE

The infrared and RF wireless LANs described above and listed in the charts all provide reliable wireless network links for this type of network extension—with the exception of the "lost" nodes discussed above.

DESKTOP CONNECTIONS

To this point we have discussed wireless connections that are designed to provide portable computer users access to a Local Area Network as though they were connected by use of a standard LAN cable. Once the connection is made, the user becomes a "node" on the system. What the user sees on the portable computer screen is what would be seen from any other node connected to the network—the user is able to use network-connected e-mail and file sharing, and network printers can all be accessed.

OTHER REASONS FOR CONNECTIONS

A further reason to connect to one's own desktop computer is to be able to move files and information back and forth between a desktop system and a portable computer.

Portable computer sales have escalated to the point where they account for one in every four computer systems sold today in the United States. The primary reason for this is that portable computers are now powerful enough to allow users to duplicate the contents of their entire desktop computer on a portable system. This is an important concept that will change as wireless connectivity becomes more robust. Users need to be able to duplicate their desktop system on the portable to enable the easy transfer of files and information from the desktop to the portable and vice versa. Typically, a user will buy a file transfer program such as Traveling Software's LapLinkV or LapLink Pro and set up the two systems to communicate with each other over a cable connected to either the serial or parallel ports of the two computers.

The next step is to run the program on both machines and then use it to transfer the files from one machine to another. LapLink Pro includes a "synchronization" mode that ensures that the files in a directory—or on the entire hard disk—are identical on both machines. Users who are not computer literate are not keen about going through the set-up routine each time they need to transfer information, nor do they want to hassle with attaching cables between the two systems.

They do, however, need to be able to make sure that the files they have on their portables match the files on their desktops. Until now, products such as LapLink Pro or remote access programs that operate through dial-up modem connections have been the only available options.

A WIRELESS SOLUTION

In the third quarter of 1993, Traveling Software and National Semiconductor (NSC) announced both a technology and a business partnership based on a wireless technology developed by NSC and communications software based on Travelings' LapLink products. LapLink Wireless, called the first personal wireless docking station, was the first true shrink-wrapped solution for exchanging information between a desktop and portable computer using wireless technology.

The most important aspect of this product is not what technologies the two companies have brought together—it is the product itself. When a portable computer is brought within the range of its brother desktop PC, the two units can automatically—without user intervention—begin communicating, exchanging data, and updating and synchronizing both systems.

The user does not need to attach any cables, load any software, or launch any program. After initial installation of the system and the selection of automatic operation, the process occurs every time the two units come within range of each other. This combination of products has been designed to work together, to be easy

Figure 3-1. *LapLink Wireless*

to use, and to provide the end user with a solution to what has, for dual computer users, become one of the major drawbacks—making sure that all the information that needs to be on both machines is, and that such information is identical on both.

NSC's AirShare's tranmission range is about 30 feet between transmitters. It is not intended to be a wireless LAN, nor is it intended to be a wide area network. It is intended to connect a desktop with a portable when the owner walks back into his or her office or study and the two computers come within range.

Consider those who use a computer both at work and at home, and have two systems—a desktop at work and a portable that is carried back and forth between home and work. LapLink Wireless assures that when they leave for the day, the information they have been using on the desktop is on the portable. If a user also has a desktop computer at home, the addition of a third transmitter/receiver installed on the home desktop will update this machine as well. This user can work at the office all day and simply pick up his or her portable on the way out the door. When he or she arrives at home, all the information used at the office will be transferred to the home desktop. In the morning, this process is reversed. When returning to the office, there is no need to wonder whether the right floppy disk was remembered.

One of the most interesting things about this product is that it does not make use of any new or advanced wireless technology, nor does it even push the limits of technology. Instead, it uses technology that has been available for a long time, reducing it to a solution that is more important than the product's specifications.

National Semiconductor brought the radio transmitter/receiver technology to this partnership. Its products, AirShare radio modules, are low-powered transmitters and receivers, presently built into a small, lightweight housing somewhat larger than a PCMCIA card form factor.

AirShare is a transmitter/receiver operating in the 902–928-MHz band (below the cellular channels), under Part 15 of the FCC rules. These devices do not require a license from the FCC for their use. While most wireless LANs and other devices using this band have gone with spread spectrum technology, National realized that using standard FM (Frequency Modulation) techniques would yield the best results for this product and would keep down the cost.

Accordingly, the FM bandwidth for these units is about 250 KHz, which easily accommodates a data speed transfer rate of 115 KBps (Kilobytes per second). While this speed is slower than existing LAN and wireless LAN technologies, it is the same as that of infrared. Since most operations are completely transparent to the user anyway, this is plenty fast enough for this type of use. There is also a switch on the side of the unit so the user can select one of three radio channels on which to operate, providing flexibility to use multiple units in the same building. With a range limited to 30 feet, and a choice of three channels, it would be possible to outfit many users on the same floor without having to be too concerned about interference.

AirShare units require very low power and can run for 5 to 6 hours on a standard 9-volt battery. The unit has been designed to operate using a 9-volt battery slip-on pack, power obtained from the mouse-port (pass-through cable included), or with an AC adapter (also included). The unit measures 2.25 inches wide by 3.5 inches tall by 0.5 inches thick. It can be attached to a computer with Velcro for ease of use, or it can sit in the 9-volt battery unit for desktop use.

On the software side of the product, Traveling Software provides LapLink Remote Access which provides remote drive and printer sharing between two PCs in Windows or DOS, as well as LapLink Wireless. LapLink Wireless provides the new Synchro Plus software that runs in the background in Windows and provides automatic synchronizations whenever a connection is detected. (We dubbed this "Lap*Blink!*")

A user can determine which files, sub-directories, or directories are to be synchronized, or let the system determine which files have been changed or updated and synchronize them if the same file has been changed on both systems.

LapLink Wireless can be password-protected to provide another level of protection and to prevent files from being inadvertently loaded on someone else's similarly-equipped system.

Both companies present this product as a low-cost method of seamlessly and automatically moving information from one computer to another. They are also quick to point out that since the devices are radio transmitters there can be some interference caused to portable computers if the transmitter is attached to the back of the case behind the screen. This is especially true for portables that use active addressing color screens, since each pixel on the screen has a transistor behind it. This is not a major problem—a little experimentation with unit placement can easily cure any interference problems that might be encountered.

The second caveat is that not all mouse ports provide DC power that can be used to run the transmitter/receiver unit. The NEC Versa UltraLite, for example, does not provide a DC power lead on the mouse connector. While Traveling Software and National Semiconductor have provided other ways to power the unit, using the power supplied on a mouse port is the most convenient. Potential users need to determine if DC power is really available with their particular unit.

MOTOROLA'S ALTAIR

For those who do not want to gamble with RF interference or who have a need for higher speed wireless connections, Motorola offers an RF wireless LAN referred to as Altair. This LAN system operates in the 18 GHz range and is licensed by the FCC. Motorola provides several versions from the basic Altair Plus system introduced in March of 1992, to Altair VistaPoint for building-to-building LAN communications, to the Altair Plus II system announced in November of 1992.

The speed of the Altair system competes directly with its wired cousins—up to 10 MBps—and the system is limited by license to five channels for a 35-mile diameter area. Each "channel" makes use of two 10 MHz-wide frequencies per system (which is the reason for the higher speed capability—speed is dependent on bandwidth).

The Altair system requires two transceivers—one located at the node and one connected directly to Ethernet servers and hubs. Altair supports 10Base-T, 10Base2, and 10Base5 network systems (Ethernet twisted pair cabling), as well as SNMP MIB II and Extended MIB, permitting full network management.

The advantages of the Altair system are, first, that Motorola controls the licenses and can better manage the interference potential. Second, the through-put of the system is about 5.7 MBps, approaching true LAN speeds.

Motorola designed this system to be a complete LAN solution. System parameters permit up to 50 devices per microcell and each microcell will typically cover 5,000 square feet when including 2–3 walls, and up to 50,000 square feet in open access.

WIRELESS LANS

There are many choices for wireless LANs. The best solutions at the moment are for point-to-point node placement, but there is hardware and software available for multiple node connections as well.

The Photonics infrared system has the best chance of becoming the *de facto* wide-angle infrared standard because Photonics is currently working with a number of major portable computer vendors. These vendors will be offering the Photonics system as an option with a number of portables. Users who plan to use infrared should make sure that the systems are compatible. The Photonics system uses a combination of defused (wide angle) and point-to-point infrared beams.

Point-to-point infrared is easier and less expensive to implement in portable computers than is wide angle infrared. Hewlett-Packard (HP) builds point-to-point capabilities into its two handheld computers (the HP 95LX and 100LX), as well as its sub-notebook systems (OmniBook 300 and 400). HP's PC division also includes infrared capabilities in its desktop products.

The Apple Newton MessagePad also has a built-in infrared transmitter/receiver. When is was first released, however, communications were possible only between Newton units—desktop computers or printers with Newton's infrared capability were not available, and the Apple infrared system was not compatible with the HP system. Tandy and Casio had the same incompatibility problem with their Zoomer PDA—two Zoomers can "talk" to each other over infrared, but that infrared is not compatible with any other on the market.

THE IRDA

In the summer of 1993, the Infrared Data Association (IrDA) was formed and aligned with the Portable Computer and Communications Association (PCCA). This non-profit organization's charter is to develop an industry-wide infrared standard for point-to-point communications. More than fifty companies have joined the IrDA, and they are actively working toward such standardization.

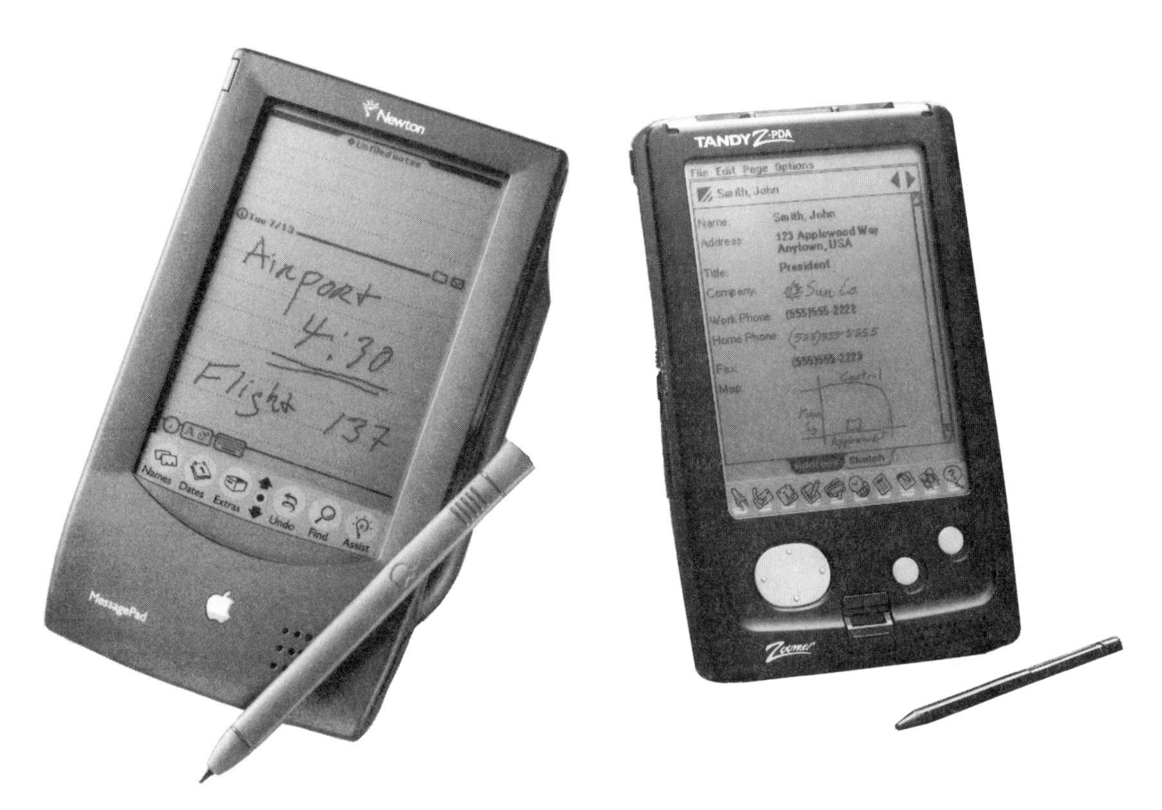

Figure 3-2. *The Apple Newton MessagePad and Tandy Zoomer PDA*

The IrDA planned to announce its infrared standard in March of 1994 at The Hanover Fair (CeBit) in Germany. Products making use of this new infrared standard are to be available by the second quarter of 1994.

When buying products that make use of point-to-point infrared communications links, users should verify that the devices they plan to buy comply with the IrDA standard to ensure that they can be used to communicate not just with like devices, but also with other handhelds, printers, and desktop computers.

Users who want to make use of wireless communications for data within a building or campus as part of an existing network will discover that it is easy to do so. Going beyond this form of in-house data communications capability is, for the most part, still in the developmental stages. A lot of activity is being concentrated on wireless PBXs (telephone switching systems), and some companies are readying total wireless solutions for use within the business environment. Still others are experimenting with the incorporation of video, audio, and data on the same wireless systems.

Some of these efforts will gain momentum once the FCC has worked out all the wrinkles for the new 1.8-GHz unlicensed portion of the spectrum. Further, once the licensed portion of the PCS band has been auctioned off and systems are set up, the combination of on-campus, local, and wide-area systems will provide more interesting connection possibilities.

Until then, though, wireless LANs can be used within a building, between buildings, and throughout a campus. Wireless LANs are here and available now, but users should be sure the costs can be recovered over the short term. Over the long term, many more options will be available from which to choose.

Company	Product	Infrared	Radio	Frequency	License Required	Novell-Compatible
California Microwave	RadioLink	No	Yes	902–928 MHz	No	Yes
InfraLAN Technologies	InfraLAN	Yes	No	not applicable	No	Yes
Motorola	Altair	No	Yes	18 GHz	Yes	Yes
NCR	WaveLAN	No	Yes	902–928 MHz	No	Yes
O'Neill Communi-cations	LAWN	No	Yes	902–928 MHz	No	Yes
Photonics	Wide Area and Point-to-Point products	Yes	No	not applicable	No	Yes
Proxim, Inc.	RangeLAN	No	Yes	902–928 MHz	No	Yes
Traveling Software and National Semiconductor	AirShare	No	Yes	902–928 MHz	No	Portable to desktop
Windata, Inc.	FreePort	No	Yes	902–928 MHz	No	Yes

Figure 3-3. *Wireless LAN Products*

IN-HOUSE SYSTEMS

The one area we have not covered in our quest for in-house wireless communications is that of custom systems designed for specific purposes. For example, in the new Tandy Amazing Universe stores, employees carry GRiDPad pen-based portable computers with built-in RF communications capabilities strapped to their wrists. These pads are a wireless component of the store's computer system, and they enable employees to provide complete service from the sales floor. They can check inventory, update customers' profiles, and place orders. Customers proceed to a check-out line to pay for their orders, and their purchases are ready for pick-up as they proceed through the warehouse.

Other vertical market applications for wireless connectivity include inventory control in warehouses, order-taking in restaurants, truck routing, and, of course, the grocery store and other point-of-sale merchants. Systems for these types of operations are up and running in almost every industry segment where mobile workers are gathering or using data. For example, the Hertz and Avis check-in people who meet customers in the rental return lane. Before they can take their bags out of the trunk, they are being presented with a bill—completed via a wireless terminal and then sent to a printer that the employee carries.

Our focus is the ability to connect—and make use of—generic computing devices in an office environment without the need for network cabling. Over the next few years, the types of devices presently used for specific vertical markets—with very specific software developed for a specific company or industry—will give way to machines that not only provide these functions, but can also be used to connect the user to a variety of other information sources and data via the wireless connection.

4

Wireless Wide Area Networks

Wide Area Network (WAN) wireless communications are not nearly as far along in the development cycle as are wireless Local Area Network systems. That is not to say that LAN wireless systems are robust and are being used extensively in all types of businesses—there is still a long way to go. Wireless WANs lag behind LAN development not only because of a lack of frequencies, but also because of the lack of an infrastructure required to provide even the most basic of services.

Within the last five or six years people have discovered they could take a portable computer on a trip and, with the use of a modem, access their office or other e-mail systems. But this is still not an easy thing to do—even over standard telephone lines. Part of the reason that this is not as easy as one might expect is that telephone systems were built for voice information. The idea of "plugging into" the phone with other devices was not thought about or planned for until the past decade.

Of course, the disadvantage of making use of the existing phone circuits is that the user must be near one—and then, of course, that the user must have access to the circuits. Further, users may have to work with both the software and hardware to achieve a viable configuration for the new location. Also, once connected, the user may discover that the local system is not compatible with his or her hardware.

WIRELESS TO THE RESCUE

An area-wide wireless communications system offers many advantages over having to access wireline systems (area-wide being defined as that area in which the system must operate—whether ten blocks, ten cities, an entire state, the nation, or the world). This is one reason mobile cellular and portable cellular phones have sold so well. The portable phone, particularly, belongs to the user who can use it almost anywhere, for any purpose. People have flocked to this form of mobile communications. It is easy to use, relatively inexpensive, and can be used almost anywhere in the nation or the world.

Over time cellular, as we know it, will be expanded to include nationwide and even worldwide service by use of satellite systems. The processing of data will certainly be a part of this increase in service and capability. For now, however, making use of the cellular system to send and receive data appears to be a less than adequate solution. Presently, there are too many variables. There are several new technologies emerging that may make cellular work for data, but to start with a system designed for analog voice and then to seek to superimpose data does not seem to be the best tactic for the development of a foolproof and reliable solution.

Once implemented, the new Personal Communications Service (PCS) may provide a nationwide and worldwide voice and data system. PCS systems will be built from the ground up as digital systems—making both the voice and data portions of the system more robust. Unfortunately, the FCC has decided not to issue nationwide licenses for these services. Thus, we may find ourselves back in the same situation as with analog cellular—there is no one system that provides instant access under all conditions.

BACK TO THE PRESENT

Currently, implementing a mobile computing system on a regional or nationwide basis is possible. It may not be as "seamless" as users have been promised (that is in the future), but it will work. Combined with analog cellular, such systems will enable a company to provide both voice and data communications that work well enough that the general user will not be frustrated.

Frustration is a major factor in the area of technology failure. If people expect too much because the technologists have "sold" them too much, the actual results translate into something less than expected.

CATEGORIZING WIRELESS WANS

Wireless WANs are at about the same point as wired modem communications were five years ago. That is, users want to implement wireless communications, and they really need to be able to move voice and data to any location in an area or region, it can be done. In many cases, though, this technology may still be too complex for users who are not computer-literate. Also, it may prove frustrating for those who have not taken the time to work with (and understand) wired modem communications. However, with a little education and with some trial and error, wireless WANs are capable of providing a good return on investment and a measurable increase in productivity.

TYPES OF WANS

This discussion of WANs addresses not only what is available and possible for two-way systems, but also what is available for one-way systems. For now, one-way systems may be considered one-way WANs since it is possible to move data from a given location to any number of units in the field using wireless technology.

TWO-WAY WANS

Any two-way radio system is capable of being used for data transmission. Before there was such a thing as "wireless computing," many two-way radio systems were passing data and voice between mobile units and base locations. One of the first uses for this type of combined system was in police departments where a computer terminal was installed in a police car and data messages were sent to it. Such messages can be in the form of requests for services or license checks. The object is to keep as much voice traffic off the congested channels as possible and yet to be able to provide the field units with timely information.

Fire service offers another example of mixed voice and data systems. In some locations, the first alert given to a fire department is a radio signal that is decoded at the fire station and sets off an alarm. This alarm is then followed by a voice message giving the location of the fire. In some systems—such as the one that has been used in Phoenix, Arizona, for a number of years—the dispatch address is also sent to mobile data terminals in the fire vehicles.

En route to the location, the fire trucks are given route and condition updates with the use of data modems. If the fire location is a factory or public building, the firemen can even receive a building plan and indications of any hazardous materials stored there. This information is designed to save time and to help the fire department know what to expect when it arrives on the scene. This system has saved lives and property on a number of occasions.

The data systems most familiar to us are those used by Federal Express and United Parcel Service (UPS). These systems were field-tested and installed because of the need to track parcels no matter where they are in the delivery cycle. The UPS package tracking system makes use of the ARDIS and cellular networks, and the Federal Express system (a private system) uses a combination of satellite, microwave, and radio technologies for voice and data links.

WHAT IS ARDIS?

One definition for ARDIS is furnished by the Microsoft Word spelling dictionary which proposes "RADIOS" when it comes upon "ARDIS" in the text of an article. However, it is really an existing, specific data radio system jointly developed by IBM and Motorola.

Years before wireless data became the "in" thing, IBM combined forces with Motorola and built a wireless data network for its field service personnel. Motorola installed a nationwide backbone data-only radio system and, in conjunction with IBM, furnished wireless data terminals for the IBM service force. The system has proven to be very successful, saving IBM countless dollars and time. At the same time, it has validated the concept of remote, wireless computing.

This first implementation was designed to provide access from the field to existing computer systems in each of the IBM field offices—sending and receiving service call data, requests for service and parts, and billing information. Over time, ARDIS has expanded its system so that it now covers more than 100 cities throughout the United States. It is being used to provide direct data connections between field and office work forces.

Although the majority of ARDIS users require only a connection between a captive field force and their own office-based computer system, ARDIS has also begun to market the system for individual users. Motorola introduced its InfoTac radio/modem combination and signed a deal with RadioMail (see below) putting it into direct competition with RAM Mobile Data.

The ARDIS system uses a single radio channel throughout the United States. This nationwide frequency is being supplemented with other channels in specific cities as demand for the service grows.

RAM MOBILE DATA

The only competitor to ARDIS is RAM Mobile Data. While the ARDIS system has been in operation since the mid-1980s, the RAM system in the United States was officially turned on in December of 1992. By mid-1993 it had more than 100 cities on-line (covering about 90 percent of the U.S. population). The system is based on the Mobitex network pioneered by Ericsson in cooperation with Sweden's National Communications Authority.

At present, Mobitex networks are operating in the United States, Canada, United Kingdom, Sweden, Norway, and Finland. Additional systems are under construction in Australia and continental Europe. Although Ericsson developed the protocols for the Mobitex system, there is an organization called the Mobitex Operations Association (MOA)—consisting of representatives from each network operator—that follows an open-protocol policy. MOA publishes full protocol specifications for radio modems and terminals. Equipment that works on the RAM system is being offered by Ericsson, Motorola, and many other manufacturers. End users have a wide range of equipment to choose from.

The main difference between RAM and ARDIS is that the RAM system makes use of a radio technology referred to as "trunking." Trunking involves a number of different radio base stations on a number of different radio channels. Channels are automatically switched as service is requested by end users. The advantage of this type of system is that it can handle more traffic per geographic area than a single-channel system. Experts seem to agree that between 500 and 1,000 users can be supported on a single packet data channel per geographic area. While ARDIS has one channel per geographic area and is adding channels as demand increases, RAM has 10 to 30 channels per area.

Until recently, most of the RAM and ARDIS users were corporations with a need for internal wireless data systems. These are used for package delivery (such as the UPS system), field service, or field sales applications. In November of 1993, RAM announced it had also signed a deal with RadioMail and was in a position to provide extended or wireless e-mail service to individuals as well as to corporations.

With the advent of these "personal" data services, wireless data capability has come to the individual. User costs are based on a per-packet fee or monthly charge. The ARDIS system, at present, runs about $.15–.17 per packet and RAM's system is priced at between $.03–.05 per packet. (A packet is a "packaged" stream of data that contains sender and recipient information, a complete message or a portion of a message, and information used to provide error correction.)

HARDWARE AND SOFTWARE

Both the Ericsson Mobidem and the Motorola InfoTAC are a combination of radio transmitter-receiver units and a data-compatible modem. Both units send and receive data at speeds between 9,600 and 19,200 baud and both can be plugged into many different computers. RadioMail provides software that runs on any standard DOS machine, the HP 95LX and 100LX handheld, Macintosh, and even pen-based systems running pen-enabled DOS from CIC or GRiD. (The RadioMail interface for pen-based systems was written by PenPal using its software developer tools.)

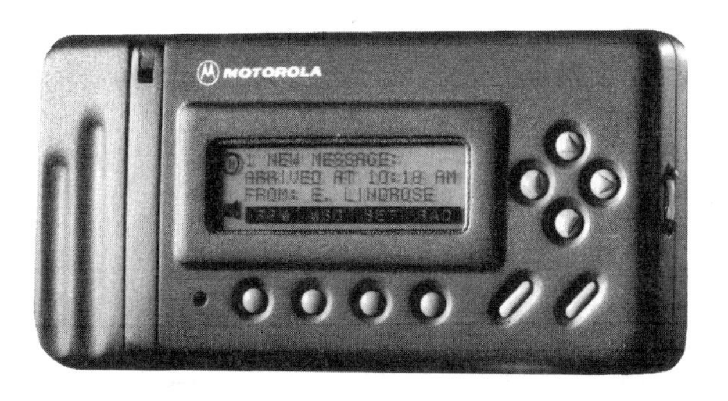

***Figure 4-1.** The Motorola InfoTAC*

USING A MODEM/RADIO

The Ericsson Mobidem and the Motorola InfoTAC are designed to plug into any number of portable computers. They interface through the serial port on most systems and cables are available for the HP 95LX and 100LX, any standard DOS laptop or notebook, and units such as the GRiDPad and the Fujitsu PoqetPad. Macintosh and Windows versions of the software are either already available or will be available within the next few months.

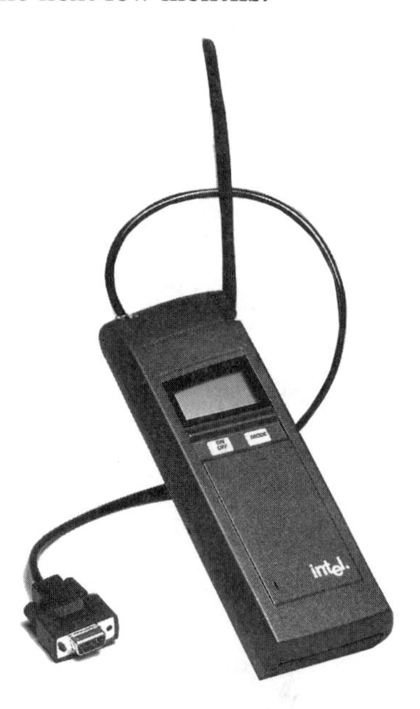

Figure 4-2. *Ericsson's Wireless Modem, Marketed by Intel*

In order to implement wireless communications on an individual basis from a portable computer, users will need to purchase one of the above-mentioned radio/ modem devices, load the RadioMail software into your portable, and sign up for RadioMail service. Once service has been established, the system works just as one would expect any e-mail system to function.

When the two devices (the portable and the radio/modem) are turned on and the software is loaded, the system (either ARDIS or RAM) is accessed and a packet of data is sent to the RadioMail gateway over the wireless link. Once the RadioMail gateway "learns" that a unit is up and operational, it queries the system for any mail that might be waiting for that address and then downloads it to the system.

Once the mail is on the system, the user can read each message, respond to it, store it, and delete it. The user interface for DOS is easy to learn—function key commands are used to read, save, store, and delete messages. RadioMail provides all the user interface and protocol conversions required for access to the Internet, MCI Mail, AT&T Mail, the user's internal mail system, and many other services.

RADIOMAIL

The best way to understand what RadioMail is all about is to consider it a repository for information. RadioMail serves e-mail systems, NewsFactory, and RadioFax. Over time, it will provide access to sports, weather, and many other databases. Further, users who want to have their MCI or AT&T mail "forwarded" to their RadioMail address can set up either system to "auto-forward" their messages to RadioMail. At present, this means they will receive *all* of the mail sent to an MCI or AT&T mail address, but in the future they will be able to access their mail box, request a list of pending messages, and select the ones they want to be sent to them over RadioMail.

The beauty of the RadioMail system is that it provides *all* of the protocol conversions necessary to move data from one system to another. The end user sees nothing but a common e-mail interface. RadioMail has solved the problem of differences in protocols—it takes virtually any service and converts the data to a RadioMail standard and sends it down the radio pipe. Going the other way, the end user has to do nothing more than key in the address and the message and press the "send" key. The RadioMail system performs any required reformatting internally—without end-user intervention. RadioMail service is available on both the ARDIS and RAM Mobile Data networks.

NEW PARADIGMS OR OLD?

The wireless LANs we have discussed thus far offer users the same view of a network as they have when they sit in front of a desktop computer. Wide-area options, however, use proprietary techniques for sending and receiving packets of data. For this reason, the first applications used over such wireless networks required extensive modifications for them to function properly. One of the main attractions of the RadioMail system is that it converts the various protocols used by other networks and provides the connection between networks. End users do not have to participate in this activity. They simply need to know the e-mail addresses of the people to whom they wish to send e-mail—the rest is handled by RadioMail.

But what options are open to a user who wants to connect to a corporate e-mail system using a wireless data network? One can contract with RadioMail, RAM, or ARDIS and provide the selected provider with a connection to the e-mail system, but then the user must have special software to handle the connection. Depending on how it is implemented, the results could be that the user has mail in two different places, or has mail forwarded from one place to the other without any indication that a message has already been received and read.

The biggest hitch in such a plan is that the Information Services (IS) folks in most companies are already busy trying to provide good, solid computing services to the entire company. They generally do not have the time or the budget to experiment with wireless e-mail and system connectivity.

Most of these same IS departments already support wired modem connections to their hosts or Local Area Networks, however. If a compatible method of providing wireless connections could be defined, the IS director would probably be more willing to experiment with wireless connectivity. Further, if an existing technology were modified so that the wireless connection "looked" like a dial-up modem connection, the software running on both ends of the system would not have to be modified much (if at all).

To this end, the Wireless AT Modem was developed by Ericsson. This device was chosen by Intel for its private-label wireless connection strategy. Ericsson developed this modem with the cooperation of RAM Mobile Data and Research In Motion (RIM). The idea is simple enough: Make wireless transmissions that are packetized, and connections that resemble direct-connect wired modem connections—the systems will talk to each other and messages will be passed between them.

There were a few more obstacles to overcome, but soon the "Mobidem" was ready. Agreements were reached with Lotus for modifications to cc:Mail, similar modifications were made to Microsoft's mail system, and even AT&T got into the picture with its EasyLink public e-mail system.

The main issues to be addressed were the fact that when a modem connects over a telephone line to a system, the circuit (or line) remains up and running during the entire exchange of information. If one end or the other does not "see" the phone connection, the call is terminated and the user is told to try again.

In a wireless data system that is based on segmenting a message, sending it as a number of data packets—and putting it back together on the other end—there is no direct connection for the modem to "see." Unless some way is found for the packet connection to be seen as though it were a direct connection, the modems on one or both ends of the system will time-out, thinking that they had lost the carrier from the other machine's modem.

IMPORTANCE DOWNPLAYED

When the first Intel wireless modems were introduced, most communications companies were quite vocal in their criticism of this "compromise" solution between the wired AT modem and a packet data network. They felt that this technology returned wireless connectivity to "the dark ages." As it turns out, the use of the familiar AT modem command set for wireless modems was better received by IS departments than was expected. Software vendors found that their software offerings needed little, if any, modification to run on a wireless network, and IS departments were more comfortable using a known technology that does not disrupt their current operations.

Shortly after Intel/Ericsson announced their AT Mobidem—and RAM endorsed its use—the Portable Computer and Communications Association (PCCA, a nonprofit industry organization) undertook the task of standardizing the commands that would be used in a wireless environment. The AT Modem committee of the

PCCA was able to structure standard formats and command sets that would permit the early adoption of wireless technology.

The final specification is available, and portable computers employing this technology were to be available in the second half of 1994. The advantages of using a common AT command set for control of both wired and wireless modems is that software vendors should not have to greatly modifiy their applications for them to work over a wireless connection. Many computer users are already familiar with the set-up and operation of wired modems, and they should find very little difference in using a wireless version.

The AT Modem committee played a vital role in providing wireless standards. After numerous technical working sessions, it voted on the recommendations and they were accepted. Microsoft, IBM, Compaq, Motorola, Ericsson, and others realize that the AT modem command set is not the be-all/do-all solution, but will it play an important role in advancing the industry. Over time, this issue may diminish in importance, but while wireless connectivity is still in its infancy, standardization should attract users to existing networks.

The use of extensions to computing standards that have already been mastered by computer-literate people is perhaps the most important factor in raising their comfort level and persuading them to participate in the move to a wireless world.

GOING WIRELESS

Going wireless today is not as easy as it will be in the future, but it is not any more difficult than buying a telephone modem and learning how to deal with "dialing" into a wired system. Both the Ericsson and Motorola units are self-contained radio transmitters/receivers and can be attached to a number of different computers with a serial cable connection. Once the proper communications software is installed, the system is used much like a standard modem connection with the exception that the systems do not require dialing sequence. Instead, once the system is turned on, the portable unit checks with the wireless system and establishes the radio link.

Once the links are established, users can send and receive mail just as though they were connected to their own e-mail system with a direct hardwire connection. Also, unlike dial-up systems (over wired or cellular systems), the wireless service provides a direct connection to a single mail source, or through a service such as RadioMail, the connection is made to an information repository and preformatted information is automatically forwarded to your system.

ONE-WAY WIRELESS

There is another way to send data to field personnel—with radio pagers. Pagers come in many varieties—some provide a tone alert, followed by a voice message. A numeric read-out pager displays a phone number to be called. Some pagers receive and display alphanumeric messages. And there is a new breed of pagers designed to be used in conjunction with portable computers.

Stand-alone alphanumeric pagers can display messages up to 256 characters or more in length. They are available in most markets as local-only systems. They are also available from companies such as SkyTel, MobileComm, and EMBARC (the Motorola nationwide system) that offer nationwide coverage and, in some cases, coverage in other countries as well. Access to these pagers is through a computer system and dial-up modem. The address (pager code) and message are entered on the computer, the paging system is dialed, the information is transferred to the system, and then sent out to the specific pager—or group of pagers—called out in the address.

Paging is the most economical form of broadcasting to one or to many and can be used for notifying a field force of a company-wide meeting, reminders that reports are due at the end of the week, or any other type of sales and service advice that applies. In this form, instead of receiving merely a number to call back, information regarding the call is also transmitted. When the call-back is made, the caller (pager user) knows what the call was about and is usually prepared with the answer, saving time and, in many cases, impressing the customer with the quick and complete answer.

MESSAGING SYSTEMS

SkyTel, MobileComm, and EMBARC offer pagers that couple with computers to receive longer messages and to display them on a computer screen. The EMBARC system, for example, sends daily news snippets from *USA Today* that can be viewed by the user on his own computer screen. The SkyTel, Mobilcom, and EMBARC pagers are made by Motorola and directly interface to the HP 95LX and 100LX. These pagers can be connected with a cable to any number of other portable computers.

All these systems permit access in a number of different ways. Thus, the user is more accessible than he is with some of the two-way systems that *require* access by use of a computer or terminal.

For example, if a messaging customer has a pager that is capable of receiving both numeric and alphanumeric messages, and the system is designed for multiple types of access, someone wanting to contact the pager user can choose between a number of options.

The simplest way to access the pager is to dial an access number (usually a local no-charge number, or a toll-free 800 number). When the system answers, the Personal Identification Number (PIN) is entered, followed by the number to be called. The pager user will then be paged and the number entered will be displayed.

Another way to access the system is to call an operator, identify the user you want to access (again using the PIN), and dictate a voice message to the operator. The message will be keyed into a computer and sent to the pager. The message will be displayed as it was dictated.

The final, most effective way to reach a person is to use computer software and a modem installed on a PC. The message is typed and sent via the modem to the paging system. The pager user sees exactly what was typed.

A real advantage to paging or messaging systems is that a user can send the same message to a number of different people without having to re-key it into the system and call each pager—a pricing update can be sent to fifty field reps.

Software for these pagers differs from system to system and from supplier to supplier. Software with the EMBARC system, for example, provides for display and storage of messages. The interface software provided by Hewlett-Packard for the SkyTel system also permits "dropping" the information received into the HP 95LX and 100LX Personal Information Manager (PIM).

Other companies, such as Ex Machina, have developed e-mail-to-paging software products that provide a direct tie between an in-house e-mail system and a pager/computer system. In this way, e-mail to a person who is away from the office arrives at the pager and can be displayed on the computer.

USES FOR PAGING

One-way paging is more than just half of a two-way data link. Information can be transmitted, received, and reviewed. Nevertheless, answering requires a different medium, such as a telephone response. But because the information vital to the intended party was received by that person, the action cycle is shortened. In many cases where a direct response is necessary, this can be just as effective as a full two-way data system.

Paging can be useful for the applications mentioned above, or it can be used to inform key individuals of events or agenda changes that need to be noted, but not necessarily acted upon quickly. Other uses include notification of changes in information, changes in stock market prices of a given set of stocks, or any other event that can be conveyed by e-mail or voice mail.

E-mail and voice mail messages are, after all, one-way communications that can be answered later, by use of some alternate form of communication. There are many times when one-way communication will fulfill the prime objective of conveying information to the person or people who need to act upon it in a timely and inexpensive manner.

Using the newer alphanumeric paging systems provided by SkyTel, Mobile-Comm, and EMBARC, a user with a messaging receiver attached to his computer can receive large amounts of information that can be stored for access later.

One of the most important developments in the wireless messaging area is the introduction by Motorola, Wireless Access, RC&T, and others of PCMCIA card form factor messaging receivers. Such receivers are designed to be inserted directly into handheld or notebook computers that have a PCMCIA slot. Not only can they receive long messages and information over a paging network—if the proper software is integrated into the computer—a price list update, for example, could be automaticially sent and received, and placed in the proper file in the computer. Field sales personnel would not be required to take any action—their price lists would be constantly updated.

5

Hardware Challenges

The computer vendor community has recognized that wireless communications will play an important role in the success (or failure) of their offerings in the mobile computing market. Wireless communication capabilities will be critical not only in handheld "Personal Communicators" or "Personal Digital Assistants," but also in notebook, tablet, and palmtop computing devices.

Vendors also realize that, for the most part, they do not have the expertise to design RF and wireless modem modules for inclusion in their products. Nor is it clear which of the existing (or promised) data systems need to be included in their products.

TO GO WIRELESS OR NOT

Some vendors believe they need do nothing more than provide a PCMCIA Type II slot in their units. They will let others develop communications capabilities into a PCMCIA card form factor. (The PCMCIA organization has defined Type I, II, and III compatible slots for credit-card sized devices to be used in conjunction with portable computers.) Some vendors have decided not to worry about wireless communications at all. They are providing serial port connections and are leaving communications to the communications specialists.

The entire issue of adding infrared and RF devices into or onto portable computers is confusing. For years, computer engineers have been told by the FCC to design their products to minimize radio frequency leakage by confining RF within the unit. Now they are being told by their marketing and sales departments to develop products that are RF-capable.

ENGINEERING CHALLENGES

The challenges involved in combining computers and radio frequency devices should not be considered trivial. A computer generates radio spectrum noise that can seriously impact the performance of a receiver located near the computer. Of course, this can be worse if the receiver becomes *part* of the computer. The radio frequency transmitter's output can erase RAM (Random Access Memory) and cause interference to the internal workings of a computer!

Many variables must be taken into account in the design of RF and computer products that will work together. Just because a computer has been optimized to work with one radio system does not necessarily mean it will work with systems that use different frequencies or channels. Even within the same service, specific frequencies may cause problems while other frequencies work well.

For example, cellular telephone systems make use of a minimum of 666 separate channels (832 channels for mixed analog and digital systems). This pool of frequencies is divided into two pools, one for each of two service providers in each area. When a cellular phone is used, it is switched from frequency to frequency, depending on which cell it is using and on which channel the master computer decides it should be.

Since these frequencies occupy a large amount of radio spectrum, a computer that will be using cellular radio for communications has to be designed to be immune to interference from these frequencies. Further, if the system is intended to work with RAM or ARDIS, it must be designed to be "quiet" over the 30 channel range of the RAM system and the multiple channels (depending on the area) used by ARDIS.

If the unit is intended to work with the unlicensed frequencies allocated for in-house wireless communications, it must also be optimized for the 902–928-MHz band or the 1.4-GHz band. In the future, units will also need to operate in the 1.8–2.4-GHz band that will become the PCS band.

ADDING PAGERS

Paging systems in the United States use a number of different frequencies including the 30-MHz, 150-MHz, 450-MHz, 800-MHz, and 900-MHz ranges. New systems may even use FM broadcast sub-carrier channels located in the 88–108-MHz range.

Paging receivers (as currently designed) are neither as sensitive nor as immune to adjacent channel interference as radio receivers designed for use in two-way radio systems. Pagers typically worn on a person's belt or carried in a purse respond to a digital code transmitted by a high-power transmitter. High power is used to overcome the lack of receiver sensitivity and the limitations of a primitive antenna.

Noise generated by a computer can render these receivers useless if the noise happens to be on the pager's receiving frequency. Motorola pagers that have been married to the HP 95LX and 100LX for EMBARC and SkyTel services have been coupled into the computer through the serial port. Motorola's "Advanced Information Receiver"—the Newscard—uses a PCMCIA form factor with a "bubble" at the end. The bubble houses the battery (so the pager does not drain the computer's battery) as well as the antenna.

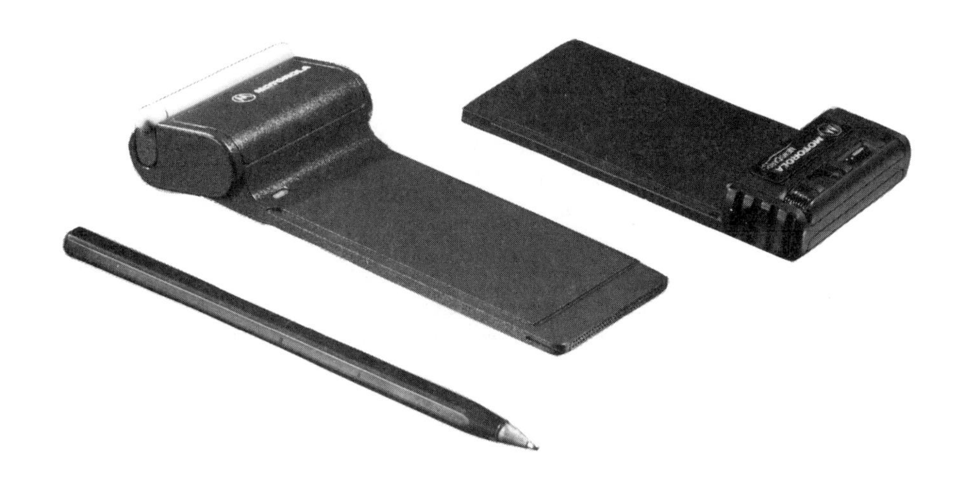

Figure 5-1. *Motorola's Wireless Modem and Newscard*

ANTENNAS

A major obstacle that must be overcome in matching a wireless device to a computer is the attachment, and placement, of the antenna. Performance of an RF device is affected more by the antenna than by any other part of the system. The better matched the antenna is to the system, the better it will work. To "match" the transmitter and receiver, the antenna must have an electrical length relationship to the radio frequency on which it is to operate.

GENERAL ANTENNA INFORMATION

For a receiver's antenna to work efficiently, there has to be an electrical relationship. If the antenna is not correctly matched to the system and to the frequency on which it is to operate, the best receiver in the world will not work properly. In the same manner, only more so, the antenna's length and placement are key to the performance of a transmitter.

Antenna efficiency is a measure of the ability of an antenna to convert RF energy from the transmitter into electromagnetic radiation. An "ideal" antenna would permit 100 percent of the RF energy to be converted into electromagnetic radiation. Practically, it is possible to achieve 50–96% efficiency. The better matched the antenna is to the frequency, the better matched it is to the impedance of the transmitter.

Impedance is stated in Ohms—a measure of resistance. According to Ohm's law, the lower the impedance, the better the energy transfer. The transmitter's impedance and the antenna must be designed to match. To understand impedance, consider a stereo amplifier that is connected to a set of speakers. Most speakers are rated at an impedance of 8 Ohms. Attaching speakers that are rated at 4 Ohms will degrade the performance of the system because of the impedance mismatch. The audio energy that does not go to the speaker actually travels in the other direction—back toward the amplifier. If the mismatch is severe enough, this returning energy can damage the final stages of an audio amplifier.

As with the speaker mismatch, when an antenna is mismatched to a transmitter, some of the energy is "reflected" back into the transmitting device where it is dissipated in the form of heat that can, obviously, damage the transmitter. The science of antenna design and construction is well understood by RF engineers, but it is still an inexact science. Just because an antenna is "matched" to its transmitter does not mean the antenna will radiate the energy efficiently.

Antennas radiate energy into the ether and must work against an earth ground of some type. An antenna designed in a lab, working against a good ground, may not exhibit the same characteristics when it is attached to a portable transmitter without the same type of ground potential.

All of these factors and more—such as antenna length, construction material, and diameter—must be taken into consideration when designing antennas that will perform well in a mobile environment. The only way to overcome poor performance of an antenna system is to design the entire communications system to compensate for poor antenna characteristics and locations.

Cellular phone systems were originally designed for use with car-installed phones using outside antennas and 3 watts of transmitter power. The first cellular sites were optimized to receive this power level (from a car-mounted phone), and were placed along major highways and heavily traveled roads. In most areas, you will find the best cellular coverage along interstate highways.

When portable phones came into vogue, the cellular systems had to be expanded to provide coverage not only for units operating at relatively high power levels but also for handheld phones that operate at 0.6 watts (600 milliwatts). Handheld phones use antennas that do not provide the same type of signal-boosting characteristics as their cellular counterparts. (Three watts is relatively high when compared to 0.6 watts. Most police and fire vehicles use radios with transmitter power from 50 to 100 watts.) To provide coverage to handheld cellular phones, new sites were added between existing sites, expanding coverage to include high business concentration areas.

Coverage in many buildings is still a problem with cellular systems, and this will continue to be the case for quite some time. The cost to "build out" cellular systems to provide adequate coverage in all the office buildings in a given city is considerable, and will only happen over time. The cellular industry has found ways around the issue of building penetration. One of these is the use of "micro-cells"—miniature cell sites placed within buildings to provide coverage within their walls.

OTHER ANTENNA CONSIDERATIONS

Radio waves travel through the air. They bounce, bend, and penetrate substances to arrive at a receiver. Some materials pass radio waves better than others. Radios within buildings that are constructed with a lot of steel will not work well, while buildings of mostly concrete and glass provide a better environment. As a user moves further away from an outside wall, however, less signal will reach his or her receiver. One of the strange quirks of antenna science is that glass is a good insulator, yet many cellular antennas are attached to a car window, passing the energy through the glass on its path between the transceiver and the antenna. The amount of loss is substantial in this case, but the efficiency of the antenna in getting the signal out into the open makes up for most of the loss caused by the glass.

POLARIZATION

Interestingly, most people treat their cellular phones very casually. Many systems installed in cars have swivel antennas that have, over time, slipped to a less than vertical plane. Users of handheld phones hold them at many different angles while talking. The best possible communications occur when the polarization of the antennas at both ends of the path is the same. Most cellular systems use vertical polarization, so keeping the antenna in a vertical position can help considerably when a user is in a fringe reception or transmission area.

Polarization became an issue in FM broadcast systems that were first installed for horizontal polarization, just as television antennas are. The receiving antennas on home FM receivers were also oriented in the horizontal plane and all was well. When FM radio went mobile, people began complaining about "fading" and poor reception. FM broadcast companies added vertical or circular polarization orientations to their antennas and, for the most part, complaints ceased.

RECEIVER SENSITIVITY

Sensitivity of a receiver is a key factor to how well it will hear a signal. But there is more to sensitivity than just making the receiver hear well. (The best way to understand receiver sensitivity is to place a single grain of sand between a finger and your thumb. You hardly feel it. However, if that same grain of sand was placed in your eye, it would feel like a boulder! Your eye is much more sensitive to the sand than your fingers are.)

There are other considerations that can affect a receiver's ability to hear. If the receiver is designed to hear specific signals on a broad range of frequencies, it is more difficult to design it to reject other signals in the same frequency range. A single frequency receiver can be tuned to be very selective in what it hears. Tunable filters are used to provide a signal acceptance that limits the receiver's ability to hear signals that are not on the exact channel for which it has been tuned.

When a receiver is designed to hear signals over many different channels, the first stages of the receiver must be broadened to accept these varied signals. In broadening the receiver "window," some sensitivity is sacrificed and there is a susceptibility to hearing signals other than those for which it was designed. While many of today's receivers are capable of receiving over the wide range necessary to work in cellular phones, SMR systems, and other multiple frequency systems, they are not as immune to interference as receivers tuned and optimized for a specific frequency.

In the past few years, engineers have made great strides in balancing the trade-offs of multi-frequency capabilities with sensitivity. But even using specially-designed tuned circuits, there are still trade-offs. If a multiple-radio-frequency system is designed properly to provide increased signal strength to the receiver, end users do not notice any effect at all—they can still receive a signal when in range. However, opening the receiver up to enable it to hear on more frequencies makes it more prone to hearing interference.

A receiver might be compared to the human ear. When a person is hearing only one speaker, understanding that one speaker is easy. However, if a person is in a crowded room with many people talking at once, he or she must focus on a specific conversation or he or she won't be able to understand it. Trying to listen to all the conversations at once will prove to be confusing, or even impossible. If the radio receiver is hearing not only the signal the user needs to hear, but others as well, the result may be useless noise.

RADIO POLLUTION

Radio spectrum is a finite resource that is heavily used and is becoming overcrowded. In large metropolitan areas, RF pollution has become a problem. Many communications suppliers rent space on tall buildings and other structures and place several different transmitters and receivers at the site. As sites get more and more crowded, the RF noise they generate increases. As the noise level increases, the receivers at the site hear less and less. More filtering must be added in front of the receivers to block unwanted signals. Each of these filters not only reduces the amount of extraneous signal, but also reduces the signal strength of the desired frequency to reach the receiver.

Generally, cellular systems are not placed at sites where many other transmitters and receivers are located, but the receiver of a cellular phone installed in a car is susceptible to receiving interference at the expense of the signal it is intended to hear. As more and more radio channels for wireless communications continue to be added, RF pollution will worsen. Therefore, optimization of the receiver and the antenna system in mobile computers is going to be a critical factor in how well they work in major metropolitan areas.

RF pollution is one of the reasons that building a single radio receiver to be usable across all our existing and soon-to-be-available wireless options is not possible. The frequencies are too far apart to allow for the possiblity of building one wireless device capable of all the wireless frequencies which will be available. Both vendors and users will have to make hardware choices based on their choices of service and frequency.

POWER SOURCES

Radio receivers are best designed to consume very little power. Most radio pagers can operate for more than a month of 24-hour-per-day, seven-days-per-week use on a single AA-1.5 volt battery. In most cases, the pager's receiver is not on during this entire time. Instead, the units are power-managed so they go into a "sleep" mode and wake up on either a timed basis, or when "hearing" the first character of a digital burst of information. They stay on only long enough to determine if the message address is for that specific unit, and they go back to sleep if the address is for another unit.

Thus, adding pager capability to most mobile computers will not place much of a drain on the computer's batteries. Despite this, Motorola's PCMCIA pager uses its own battery for power. The advantage of using a separate power source is that the user can be comfortable with the battery life of both the pager and the computer. Motorola does not risk the possiblity of negative publicity because of the pager shortening the operating time of the computer. Another advantage to using a separate power source for the pager is that sometimes radio interference can travel through the power leads within a unit, resulting in a "dirty" (noisy) power source within a computer and rendering the paging receiver useless. Additionally, the Motorola unit continues to operate even when unplugged from the computer, storing any messages received.

Once radio transmitters are added to mobile computers, battery power considerations will require even more understanding and attention.

PORTABLE COMPUTER POWER CONSUMPTION

The following chart is representative of three classes of portable computer and two different wireless options—a pager and a 2-watt wireless device operating at 800 MHz.

These figures are not real-world values and they do not take into account power management or "sleep" modes. They are representative of the design considerations that must be addressed when combining a wireless device with a mobile computer, however. As advances are made in technology, it will be possible to increase the battery life of both the mobile computer and the communications device. For now, powering the RF device directly from the same battery used for the computer is not —most of the operating time would be sacrificed.

Class of System (Generic):	Notebook	Pen	Handheld
LCD Screen, Reflective	600 mW	–	n/m
LCD Screen, Backlit	–	3.5 W	–
CPU and Electronics	2.5 W	4.74W	n/m
Modem Idle	80 mW	80 mW	n/m
Modem Connected	480 mW	480 mW	n/m
System Total	3.58 W	8.72 W	600 mW
Avail. Watt Hours	8.4 W	8.4 W	8.4 W
Total Time	2.35 hrs	0.96 hrs	14 hrs

Wireless:	Radio Tranceiver	Pager
Receive		
Idle	430 mW	200 mW
Signal	860 mW	330 mW
Transmit		
@ 600 mW	4.32 W	
@ 2 W	7.2 W	

n/m = measurement not meaningful

Figure 5-2. *Portable Computer Power Consumption*

FOR THE IMMEDIATE FUTURE

In the foreseeable future, most computer vendors will be opting for a separate power supply for the RF wireless devices they choose for their systems. Ericsson and Motorola have external devices that contain a wireless modem and transmitter/ receiver. These devices are available for use with any mobile computer which has a serial port interface and the proper software. These units will shrink in size and become available in smaller and smaller packages. Motorola is working on this project, and the Ericsson/RAM/Intel alliance should yield miniaturization of the Ericsson Mobitex unit.

Cellular phones are decreasing in size, and one can certainly envision a complete cellular phone designed and packaged in a PCMCIA 3.0 and then a 2.0 form factor. This is especially true for a phone designed for use in data-only transmissions. Omitting the voice portion of the phone will make shrinking to these sizes even easier.

The most important thing the PCMCIA form factor has provided is a target for design engineers. Everyone involved with the integration of wireless technology into portable computers is using the PCMCIA 2.0 and 3.0 specifications in their development efforts. When the marketing personnel of a communications company talk with its engineering staff, they point to the PCMCIA card as the preferred size and shape for next-generation wireless devices.

Even if the wireless device eventually will be built directly into the computer, the PCMCIA form factor still makes a good model. It is flat (only 5 mm thick for the 2.0 version and 10.5 mm thick for the 3.0 version), making it much easier to integrate into a handheld or portable computer.

Pagers will become a single chip that can be easily integrated directly onto the motherboard of a computer. Development is currently underway and plans call for the pager to be non-functional (not programmed) when the system ships from the plant. The user will choose a paging service, connect the new computer to a phone, and dial an 800 number. Once the subscription information is entered, the chip will be programmed over the phone connection.

In this way, it is possible to build a single chip that will work on virtually any paging system with virtually any paging format. The FLASH portion of the chip will accept the information sent down the phone line, and users will be able to walk out of the store with an enabled system which is ready to receive wireless data.

MARRYING THE TECHNOLOGIES

If the choices for RF services were limited to the two now available (ARDIS and RAM), the computer vendor's life would be simple. A wireless modem would be integrated directly onto the motherboard of a mobile computer (or as an add-in option, as wired modems are presently sold). The connection to the transmitter and receiver would be through this modem either to an external radio or to a radio built into a PCMCIA card.

Users would decide which of the two services to use, buy a matching radio transceiver and mobile computer, equip it with the proper software, and go on the air with their mobile computers. In the short term, this is will continue to be the best approach.

THE WIRELESS USERS

Motorola's ARDIS system has about 40,000 users of wireless mobile computing. Most of these users are connected to their own corporate computing systems and use ARDIS for specific tasks such as dispatch, sales, and repair order information. The RAM system, which currently has fewer users, is being employed for more generic computing functions.

Some users have chosen direct data connection over analog cellular systems with some success. Pagers can be married to mobile computers, and Motorola is providing such devices for its own EMBARC system and for SkyTel. The reality of wireless computing today is that users' needs can be met.

Those who have been using wireless technology within corporations for specific purposes have had a glimpse of the dream of go-anywhere, do-anything computing. The IBM service representative who carries a handheld computer and accesses the ARDIS network for dispatch and service call information has been using this type of technology for a number of years. The IBM system works, it saves time and money, and IBM is dedicated to providing the type of information needed in the field to do a good job.

The recent installation by United Parcel Service (UPS) of 55,000 cellular devices in its trucks is another example of today's wireless capabilities. In this case, UPS convinced the cellular industry to provide a central billing system and a common way of moving digital information about packages to its host computing system.

The UPS system uses data over the existing analog cellular system on a nationwide basis, but this system is very different from what is being done by most users. In this case, the cellular phone has no voice capabilities, and driver interaction is not necessary to send data. Once the driver returns to the truck and places the electronic clipboard in the special holder, the truck-mounted system dials a number. After making the connection, the data is sent over a standard cellular phone circuit.

The system works, and it works well. It works because the driver is not required to do anything to make it work. If the truck is in a bad location, or the message is not received properly, the driver is not aware that the system is re-dialing and re-attempting to make the data transfer. While there is no specific data available on the number of re-tries versus successful attempts at sending the data, UPS was able to ink a deal that means it pays *only* for completed transmissions—not for any aborted attempts. The system is not based on direct connection; the data is packetized before being sent.

Anyone who has tried to send and receive data over the existing analog cellular circuits on a direct-connect and interactive basis knows two things for sure: First, establishing contact, interacting with a remote computer, completing the data transmission, and signing off is not an easy process. Second, because any and all aborted attempts at connecting must be paid for, this is a *very* expensive process.

This is a one-of-a-kind mobile data computing system and would not exist without the persistence of those at UPS who persuaded the cellular carriers to get together and provide the central billing and coordination required to make the system work on a nationwide basis. The likelihood is that no other company will be able to duplicate this UPS accomplishment. Convincing the cellular carriers to agree on a mutual billing, and then designing a system that puts the burden of information delivery on the carrier rather than on the customer is an accomplishment that will go down in the annals of wireless computing.

DATA OVER CELLULAR

The UPS system works because of its dedicated nature, but most of those who have tried to use analog cellular circuits to send and receive computer data have been frustrated both by the number of operational problems and by the cost of the connections. Even modems that are specifically designed (according to the manufacturer) for use with analog cellular systems do not cure all the problems associated with session-based data communications over cellular connections.

Since the cellular system is already deployed throughout most of the major metropolitan areas of the United States, IBM set out to develop a method to use the cellular infrastructure to send packets of data. The first iteration of this method was called "CelluPlan."

When IBM and the cellular carriers got together to implement this technology, McCaw Cellular and a number of other carriers formed a consortium, and changed the name from CelluPlan to CDPD (Cellular Digital Packet Data). The theory of CDPD is to permit cellular operators to offer digital data capabilities over their existing cellular networks. Special equipment is added at each cell site that directs data, allocating it to channels that are not in use for voice traffic.

This is accomplished in two ways. The first is by use of what the cellular industry calls "frequency hopping." The second is by use of dedicated data channels. Both types of systems are presently being deployed by various cellular carriers. In the frequency-hopping mode, the data packets "hop" onto an unused voice channel. When that channel is put into voice operation, the packets then hop to the next unused voice channel. This technology is intended to enable cellular carriers to make use of their existing analog voice channels for data capability rather than to add additional channels.

The cellular industry is in the initial stages of deploying CDPD systems. Regional systems will be tied together to offer nationwide, seamless CDPD service. While cellular carriers are optimistic about providing CDPD coverage across the United States by the end of 1994, it is, perhaps, more realistic to expect that an additional year or so will be required to work out any technical problems to enable CDPD service to operate in a seamless manner. In the meantime, as various cities and regions continue to come on-line, CDPD may, in fact, be a viable option for those who have a need for wireless data connectivity on a regional basis. (For a complete description of CDPD and how it will be used, see Appendix A.)

THE "VISION" OF WIRELESS COMPUTING

Some amazing perceptions and plans are about to unfold. The vision is one in which in-building and wide-area systems co-exist and users move seamlessly between them. In the world of wireless voice communications, this scenario is well accepted and is being implemented in several areas of our country on an experimental basis.

The one-number, one-person voice communication concept is not simply a good theoretical model; it is practical and addresses many of the issues which must be considered in the world of wireless data. Telephone providers have learned that people are not willing to pay a premium for normal calls made while at home or within the confines of their own offices. But once they leave these areas, it is understood that there will be a cost premium associated with receiving and making calls.

Many computer and communications vendors seem to think that all users will carry their favorite handheld computer and a pocket full of PCMCIA cards. In this model, they insert the proper card for the proper type of communication, enter their offices and remove the wide area cards, and insert our local, in-house cards (either wireless or infrared). When leaving their offices to attend a meeting within the same city, users might have to replace the LAN cards with local wireless system cards. When traveling to another city, users will insert yet another card to enable them to communicate from that city.

This model does not seem practical. Users will not want to carry all these cards, and they will not want to wrestle with the issue of their location, their activities, and which system they must access. If users can have "smart" phones that know their location and which system to access, they are certainly not going to settle for anything less for their data communications.

This may be exactly what users are faced with, however. Part of the problem is that the number of communications links choices available to the end-user community will increase as new technologies and new frequencies become available. Choices available in the short term (1–2 years) will be joined by many more in the longer term (3–5 years). Even more options will become available in the future (6–9 years). The resulting confusion may be the reason that computer vendors are hoping that merely providing a PCMCIA slot or two in their systems will be an acceptable solution.

ADVANCES ON ALL FRONTS AT ONCE

At present, no one has been able to define who the major end users of all this great technology will be. Nor have the uses for wireless voice and data communications been defined. Will every traveling executive carry a personal command and control center? Will accessing all information all the time be possible? Will desk phones and desktop wired computer networks give way to wireless, walk-around, one-number access phones and an integrated data link? All this is technically possible—

but none of this will happen overnight. Additionally, converting these technical possiblities to probabilities will not happen painlessly or even inexpensively.

DEFINING THE MARKETS

How many different markets are there? How many different types of users are there? How many different kinds or classes of information do people want to be able to access? How much is needed all the time and how much on an occasional basis?

Many within the industry think wireless computing will be as popular as McDonald's burgers and that "millions and millions" will be sold. Wireless computing has been represented by many as perhaps the most important marriage of technologies since the advent of the microchip and the desktop computer. It will change the way we work, how we interact with others, and how we live our lives. AT&T and Apple have video clips of how these devices will be used in the future. These videos are exciting and fun to watch, they are probably accurate, and the depictions are certainly achievable. The question is, when, exactly, is this future?

How are end users to proceed in the meantime? How do vendors decide which products to build, how to enable them, and how to distribute them? These questions have many different answers.

Most vendors seem to be treating the evolution of wireless communications differently than they have treated the evolution of portable computers. Most portable computer vendors will not hesitate to "turn" entire new product lines several times a year. They realize that as technology moves ahead, they have to continue to develop new products. As end users demand more functionality, vendors must keep up with the competition. If these companies viewed the emergence of wireless the same way—if they were to concentrate on the present and not wait until there is a technological plateau on which to base wireless products—there would be more empowered users and, therefore, more educated feedback regarding form factors, capabilities, and uses.

COMPUTERS VS. COMMUNICATIONS

An interesting split has begun to appear between computer and communications companies. In the beginning, most computer vendors believed that the "heart" of these mobile platforms would be a computer that was married to one or more RF links over which information would be sent and received.

This model has caused many computer vendors to reach out to form partnerships with communications suppliers. Motorola, Ericsson, NEC, and other communications vendors have been working in conjunction with PC vendors in labs all over the world defining, designing, and building computing devices which are married to communications devices.

At the same time—and virtually unknown to the computer vendors—these same communications companies (and at least one computer company) have been working on "smart" phones. This started with frequency-agile systems that could tell if a user was at home, on the road, or in the office and has blossomed into a full-blown attempt to build communications devices which have the characteristic "smarts" of a computer.

The most visible evidence of this shift—or parallel path development—is found in two announcements from Motorola. Early in 1993, Motorola announced that it had licensed the Magic Cap operating system from General Magic and would be building a handheld wireless system based on Magic Cap. Then, at the CeBit show later in 1993, Motorola announced that it was licensing the Newton system from Apple for a similar product. Finally, in mid-1993, Motorola announced an agreement with Microsoft to build a Microsoft WinPad-based system as well.

SMART PHONES OR WIRELESS COMPUTERS?

If "millions and millions" of these devices are going to be sold, if everyone is "going wireless," the next question is whether everyone will be carrying smart phones or wireless computers. The next step is to define the differences between a smart phone and a wireless computer.

The most obvious difference is that a smart phone is a voice device with some data capability that uses an on-board computer to add functionality. In this model, an executive who now stands at a pay phone in an airport with a daybook, notepad, and several file folders to conduct business will use this one device to accomplish all these tasks: looking up the number, dialing the call, checking a calendar, accessing customer files, and updating information as he or she talks.

A wireless computer, on the other hand, might be defined as a fully functional mobile computer that uses a radio link to interact with other computer systems, access e-mail, track appointments, store contact and customer information, and capture notes and other important information.

In other words, until a single device is developed that combines voice, data, computing power, and all the other functions users need while away from their desks, they will have a choice of a "smart" phone with some computing capabilities or a mobile computer with some communications capabilities.

During the last half of 1993 and into 1994, several hardware vendors introduced handheld products designed to be the forerunner to what is hoped will be a personal device that all users will carry to be remotely connected.

The first of such devices, the Apple Newton MessagePad and the Tandy/Casio/AST Zoomer PDA were designed by computer companies under the premise that they are handheld computing devices that use pen input in place of a keyboard. They can be connected to many different services through communications links of various types.

However, another group of handheld devices are coming to market as well. The first of these is Simon, designed by IBM, marketed by BellSouth. Motorola has announced its intention to build such a device, and others will follow.

The difference between Simon and Newton is that the Apple Newton MessagePad is a computer with the ability to be wireless-enabled, while Simon is a "smart" cellular phone.

The distinction is important for several reasons. Most important is that Simon is designed to provide cellular voice phone users with additional capabilities. Features include an electronic daybook, address book, and to-do list. It provides the ability to send and receive faxes and to access information that is available over normal telephone channels. It is *not* intended to communicate with a user's desktop computer. It will send and receive electronic mail, but only because the technology to enable the balance of the product also provides for this capability.

The concept of Newton, on the other hand, was that it *will* be used to communicate with a user's desktop and network, and it will be able to exchange data with other computers. Further, its communications links can be anything from a wireline connection, to a pager, to any currently available wireless communications links.

The most important difference is that Simon and Newton are designed to serve different markets. Those attracted to Simon will be cellular phone users. They will buy Simon to enable them to be more productive. Their *primary* use is *not* to interact with computers, but to interact with people—by voice, fax, or some other communications link.

Newton users are computer users *first,* and cellular phone users second (if at all). They need to be able to take their computers on the road (or at least have access to the information residing in the computer) and to communicate primarily with other computing devices rather than directly with people.

As we move forward into the wireless era, it is important to understand the distinctions between the two types of products. They are designed for different users with different requirements. Those who try to lump these devices into one category may buy the wrong product for the job. Vendors bringing such products to market must also understand the differences between the markets. One of the reasons that the first generations of Newton, EO, or Zoomer have not become best-selling products is that the companies bringing these products to market did not understand who would make up the first wave of buyers. They built systems that fit their vision of the future of handheld computers, but they did not concentrate on today's computer users whose primary need is to interact with their own desktops and to send and receive e-mail messages quickly and effectively. These two activities are difficult to accomplish using a pen or stylus as an input device. Today's level of handwriting recognition, and the ability of these systems to convert handwriting to ASCII text, is not up to the task.

ONE SIZE DOES NOT FIT ALL

Total mobility does not necessarily mean that one device will meet all needs. Some will be satisfied with a smart phone, some will need the full capabilities of mainframe systems in a "handheld." Others will want to carry various pieces and parts, assembling them on an as-needed basis.

Vendors will have to offer extensive lines of products if they are to meet all the requirements of the "want-to-be-enabled" mobile user. Some users will want a tablet, some a handheld, some a full-sized notebook, and some will want a combination thereof. The ideal wireless computing system might be one that is "nested" in the desktop computer when users are at their desks. When leaving the office, they would simply push a button and a fully-functional notebook would be ejected from the desktop system, ready to be carried in a briefcase. On the airplane, this device would permit users to work on on-going projects, compose e-mail, and perform other tasks.

Upon arrival at the hotel, users would be able to send and receive the latest batch of information back to the desktop system. Upon leaving the room, they would again push an eject button and a small, handheld device would be extracted. This device would have a wireless voice and data link built-in, a personal contact list, calendar, and other information *they* decide is important to take wherever they go.

Most vendors are trying to provide only the last piece of the puzzle without the other capabilities. In addition to the vendor's unit, users would be forced to buy different types of devices and combine them themselves to achieve this vision of mobility.

VISIONS

All users associated with, or waiting to make use of, wireless mobility have a vision of what they want, how they would use it, and how much it should cost. If all visions were the same, everyone would be empowered by now, and everyone would be searching for the next technology advancement. However, since there are as many visions of these devices as there are companies and potential users, progress will be slow and no one, do-all/be-all device will be available.

Many devices are coming that will meet some, but not all, current needs. These devices will be capable enough that their adoption and use will enable the industries to better understand what the next generation of these products should look like, do, and cost. These devices will evolve over time as more vendors bring products to market.

THE MOST COMPLEX

The marriage of computing and wireless is perhaps the most complex technological challenge to be confronted since the advent of the direct-dial phone system. Users must provide the "right" information to the "right" communications carrier, to the "right" operating system, residing in the "right" hardware platform, to the "right" user. There are many different pieces to this puzzle. However, the pieces can go together in many different ways, including some which have not yet been envisioned.

6

The Mobile Computing Model

Before remote computing can become accepted practice, the ability to move information to and from mobile computing platforms will have to be greatly improved. Recognizing this, the industry uses the terms "seamless," "unassisted," and "smart" when it talks about the next level of communications and "intelligent" software.

As discussed previously, modern technology permits users to become "connected" while away from their desk or office. However, when users connect to their own computer, their own network, or an external mail system, they must use some kind of program designed to make their remote computer "look like" their desktop machine, or else employ yet another user interface, and figure out how to navigate between applications.

Thus, if users want to move a file to a mobile computer, they must be sure to have the proper application installed on their system, or they must know how to move the file in a format that can be accessed by a program that is on the system. A file cannot be sent in "native" format unless its originator knows for certain that the recipient not only has that application, but that it is the same version, as well.

For example, if a user is trying to move a word processing-document to a mobile computer, the choices are to export the file in its native format, to export it in Rich Text Format (RTF) to keep its appearance and characteristics intact, or convert it to ASCII text (in which case formatting and type face selection will be lost during the conversion process).

THE FAX

Even fax reception requires the installation of at least one, if not two, special programs on a computer. First, a user must have a fax-capable modem with fax-receiving software. Minimally, a program that will enable the computer to display and print the fax in its native graphical image form is also necessary (some fax-receive programs include this level of capability). Users who want to translate the fax image into text that can be used by a word processor will also need an Optical Character Recognition (OCR) package designed to read and convert the graphical image to text and format it for the proper word-processing program. To send changes back to the resident file, the relationship is even more complex.

Some of the most exciting advancements unfolding include the ability to send and receive faxes not in their native graphical format, but in a way that they can be viewed—and edited—just as though they were sent as a text document.

In the meantime, users must remember that a fax, by today's definition, reaches its destination in a graphical format. Converting it to a form that can be edited is not an easy matter—it is time-consuming and requires a lot of computing power.

PREPARING TO GO MOBILE

In order to "go mobile" in today's world, users must determine with whom they need to communicate, what information they wish to send and receive, and then ensure that they have installed the necessary software to accomplish their objectives.

When a user is ready to communicate over wired-modem systems or a fax modem from his or her portable, he or she must find a way to tap into the wired telephone system, load the proper program, and (oh, wait a minute!) make sure the system within the portable has switched on the modem (which is kept off in order to conserve the batteries while traveling).

MAKING A WIRED CONNECTION

Making a wired connection is far easier than it was only a few years ago, but it's not always easy. If hotel phones are equipped with RJ-11 modem jacks, it is simple.

Sometimes using a series of plugs and adapters is necessary; sometimes one must open the phone case to attach jumpers directly, or, if the phone is sealed, making the connection may not be possible at all.

Once the wiring situation has been scoped out and solved, the dial-up sequence must be "discovered" by the user. Then the connection can be made. The term "discovered" is a fair way to describe the process of determining the dialing sequence. In some locations users must dial a "9," pause, and then the 800 toll-free number. In other places, the user must dial an "8," then a "1," and then the number.

Unhappily, most communications programs do not permit easy modification of the dialing sequence, although some of the newer ones are more agile in this regard.

THE CONNECTION

Once a connection is successfully established, the user must maneuver around within the remote system that has been accessed. If a mail service such as MCI or AT&T is used, this process is fairly straightforward. Users can list their e-mail, download files they want to read, send answers through the mail service, create a fax, send a note to another electronic mail service, or even create a written response that can actually be processed and mailed by the e-mail provider through the U.S. Postal Service.

If, however, someone attaches a file to an e-mail message and the recipient does not have an application that can read and display the attachment, there is no way to receive and make use of that file. (MCI and AT&T Mail now permit attachment of documents to an e-mail message, but this capability may or may not be usable due to the complications described above).

DIRECT CONNECTION

The nature of the "connection" between a remote computing device and the information the user wishes to access is, of course, the determining factor. If a user's portable computer, using a "remote" program, is talking directly to his or her home computer, the portable accesses the desktop system and presents the user with a desktop PC screen as though he or she is sitting at the desk. Access is almost identical to sitting in front of the computer. The information available to the user is, therefore, whatever normally can be accessed from the desktop. This is the best of all possible worlds for today's computer-literate users.

If users call into a network connection using a modem with their portable, they can generally access their own e-mail account, but there is no way of connecting to their desktop computer in order to access any files across that network. All users can do is send and receive messages, and access information resident on the server.

In today's world of hard-wired connection, it is possible for users to access data, almost any data, as long as they know where it is, what form it is in, and have prepared (in advance) to be able to access it. However, in the new communications world—the world of wireless, anywhere, anytime communications—the relationships between the user and the data are more complex. If the appropriate protocols are not observed or provided, the "everyman" end user will dismiss wireless communications as not being worth the trouble.

WIRELESS TODAY

Wireless data communications today can be compared to wired data communications of a few years ago. These communications are possible, but users must know how to make the connection, know with which system they want to connect, and how, specifically, to accomplish this. Arrangements must be made in advance to assure successful communications. Consider, for example, the cellular case.

Users can connect through cellular modems in the same manner as wired modems. However, the number of errors and the problems associated with direct connection over a cellular circuit make connecting through cellular modems more difficult and more frustrating than using a standard wired connection.

Services such as RAM Mobile Data or ARDIS for data access as it exists today allows users to make arrangements for packetized connections to their own e-mail system to send and receive messages. But they are only sending and receiving messages.

With AT&T Mail and EasyLink, users can access their AT&T Mailbox, check their e-mail, and send and receive messages or faxes. With RAM or ARDIS and RadioMail, users can connect to their RadioMail address and access the Internet and, thus, access many of the other mail services—but only if they know the exact address of the individual they are trying to reach, the personal ID, and also the address of his or her mail systems. At present, none of these cases allow users to go beyond messaging.

Messaging is too complex today for many end users. However, most users think of the future in terms of their ability to access their own remote files and programs, wherever they are, as though they were at home terminal. Of course, most would also like to access data banks and resource files, copying relevant information into their own files. What will it take to make both wired and wireless data communications easy enough to be used by more than just the technically literate? A number of pieces must to come together in order to make mobile computing viable for all users. Some of these pieces are in place and some are about to become available for general use.

After the pieces are all available, they will need to be integrated into hardware and software solutions that will find their way into the field. Once this happens, mobile computing will be truly available and, therefore, truly pervasive. Before discussing the five "missing" pieces and who will be supplying them, consider what is already in place.

PORTABLE HARDWARE

As of this writing, there are no integrated mobile data devices on the market for general computing. That is to say, no generic handheld or tablet-type machines that bring together all the pieces necessary to provide true mobile computing and communications are available today. There are, however, many machines that incorporate the necessary pieces and parts to address specific vertical markets.

Most of these are made by companies such as Symbol Technologies, Husky Computer, and IBM. As stated, these machines have been designed for specific markets. The Symbol Technology units, for example, are used for bar code reading, inventory control, and other applications where a tie-in to an in-house system is required. IBM's first units were designed to provide mobile, wireless access by IBM service personnel to what has become the ARDIS network.

Another example of an integrated unit approach—one that provides specialized services to a specific group of people—is Hansen. This company's products are used for such purposes as ordering food in a restaurant. Its systems include a computing device, wireless data modem, wireless transmitter and receiver (transceiver), and specialized software to permit a single type of access to a single type of system. Obviously, the most attractive aspect of such a capability is its ease of use—in terms of its wireless portability.

TODAY'S WIRELESS E-MAIL

In many respects, wireless e-mail appears to be the most promising general-purpose offering, even though it falls short of providing the user with true desktop capability. Necessary components include a general purpose computer (laptop, tablet, or handheld), coupled with an external RF modem and wireless transceiver. Of course, specialized software is also required, but this software is designed to complete the communications link, not necessarily to provide access to specific data or applications.

Over time, computers, modems, and transceivers will be combined into a single package that can be carried and set up easily and will come closer to true desktop capability.

What are the hurdles that need to be overcome before this goal can be reached? Five missing pieces are discussed below, but there are still questions regarding which radio frequencies will be used, how they will be configured, and, of course, the protocols required for the system to operate in both a wired and wireless mode. Until these questions are resolved, computer users and vendors will opt for the types of systems available today, systems that make use of external RF devices and modems and are, therefore, not tied to any one service or set of frequencies. If RAM and ARDIS were the only two choices on the horizon, users would soon be seeing more devices that incorporate all the pieces and parts required for successful, easy-to-use wireless mobile computing.

The trouble is that there are at least two forms of data transmission over existing analog cellular systems: standard analog data and CDPD. Also, possibly as many as five to eleven new service providers are on the horizon. These providers would require new and, as yet, unallocated PCS radio channels, not to mention numerous SMR services such as NexTel or Mtel's NWN. Additionally, at least six different satellite data systems are currently "on the drawing boards." Under these conditions of uncertainty, designing and implementing a single portable device that will provide generic computing capabilities and to any one or any group of these RF data links is not practical.

MIXING LANS AND WANS

These wide-area systems do not take into account the possible combinations of wireless Local Area Networks. Such networks can include RF links in the 902–928-MHz range, as well as in the 2.4-GHz range, and the new channels set aside by the FCC for use within buildings and on campuses.

HOW CLOSE IS THE FUTURE?

The FCC is working on finalizing and auctioning the PCS allocations. Others are working on how to move the existing microwave users off these channels. RAM and ARDIS continue to roll out their own systems, and the cellular developers are working fast and furiously to implement a combination of CDPD and analog data.

It will probably be several years before a computer vendor is able to determine which of the RF technologies to include in a generic wireless computer. In fact, there may not be any clear indication of the "right" types of RF devices to include inside a box for many years to come.

How many devices will be purchased with paging builtin but with access to a system by use of add-on RF? How will users view the options of being able to buy a system that will work in a wireless RF environment in their own building, city, or state, but that will not function in other parts of the country?

How many users will require devices that will have to operate not only in the United States, but also in Europe, Asia, and the rest of the world? The transmission frequencies assigned to RAM in England are in the 430–450-MHz range and the U.S. RAM network operates in the 850–900-MHz frequency band. The systems are identical, but the radio frequencies are not compatible. This situation is like trying to receive broadcasts on a standard AM radio in a country that only offers FM radio broadcasts—the two systems are on entirely different bands and a single radio is not capable of covering both.

THE AM/FM EXAMPLE

The example of the differences between the standard AM and FM broadcast bands may not make the point since most portable radios include both sets of radio frequencies. Most people do not realize that they are actually changing not only radio channels, but also changing radio bands and reception techniques when they move from an AM broadcast station to an FM station.

The handheld AM/FM combination radio uses two different receivers, although the combination units do make use of common segments of the radio, such as amplifiers and power supplies. Still, it would not be technically possible for a single receiver unit to be built to receive both AM and FM broadcasts—just as it is not possible to build a single radio transmitter and receiver that will be capable of sending and receiving data or voice and data over all the frequencies that are or will be authorized for such services.

PUTTING IT ALL TOGETHER

When computer types get together and diagram a wireless computing system, they tend to do so in terms of the "OSI" model—linking the seven layers of a standard network architecture to those required in wireless communications (Figure 6-1).

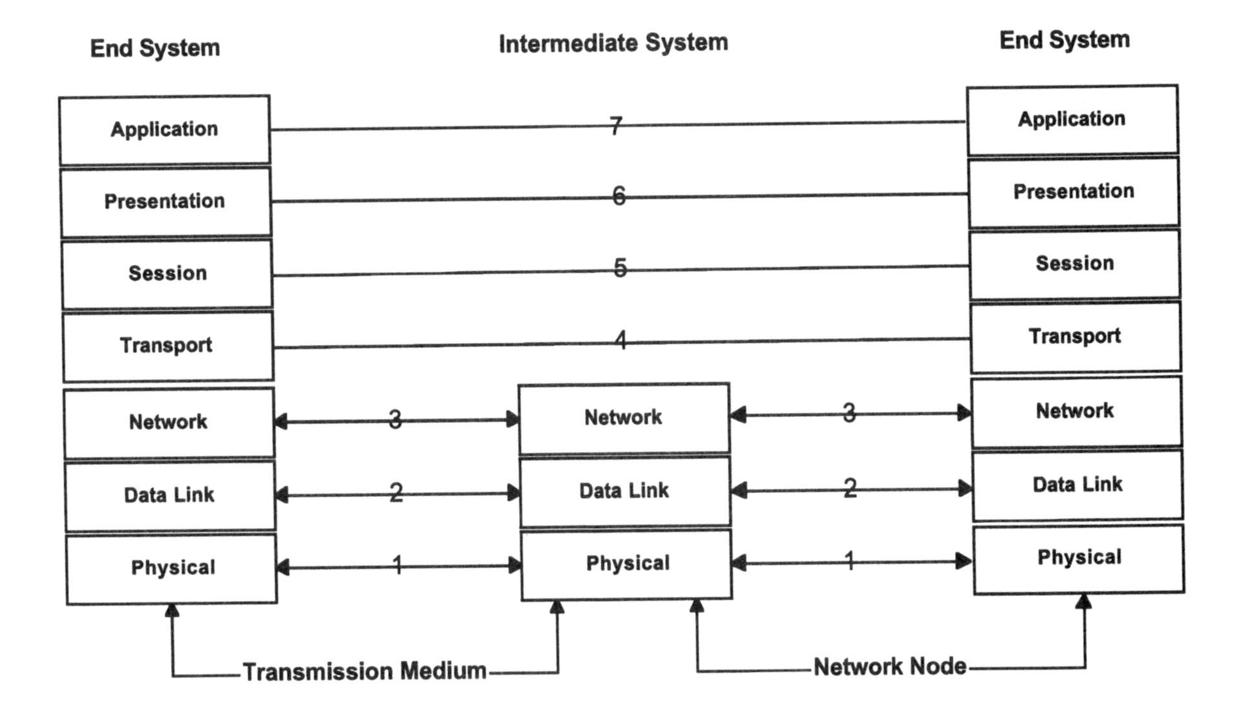

Figure 6.1 The Network OSI Reference Model

But this does not address the rest of the problem—the interaction end users have with their own computers and with the computer or information source connected to the other end of the communications pipe.

It is this interaction level that is the key to end-user community acceptance of mobile computing, and it is this interaction level that has not yet been properly addressed. To networking computer experts, the need to go beyond the first three layers of the OSI model does not seem necessary. After all, they work with this type of systems diagram all the time and if they can provide users with all the components in this model, someone, somewhere else, will take care of the user interface and interaction (levels six and seven of the OSI model).

WIRELESS

Mobile wireless systems are more than just the hardware, software, communications channels, and protocols that provide inter-system communications. The wireless connection must be able to connect to any number of different types of systems—those of information providers, e-mail networks (both local and wide-area), providers of specialized and occasional information, and to databases for retrieving and interacting with information.

Once information is received by end users, they must be able to "translate" the information into the desired format, which is not necessarily in the form in which it was sent. If a user receives a text file from an associate that must be read, marked with comments, and returned, the information must be in a form that can be viewed and edited in whatever text processor is resident on his or her system. As mentioned earlier, if a fax is received, the user should have the capability to retrieve it and view it in its native graphics form, but he or she should have the option or importing it into a document for use (requiring an OCR step either before it is sent or within the mobile computer).

Also of importance are the "snippets" of data which will be received and must be moved to somewhere else within the device. For example, if a user were to receive an e-mail message about a new appointment and it contained the name of the person, company, date, time, and location of the appointment and a map or written directions, he or she would want to store the information in the proper place within the calendar, and store the directions in a memo file linked to the calendar entry for easy retrieval.

At the same time, if this new appointment conflicted with an existing appointment, the system should be "smart" enough to let the user know that a conflict exists that must be resolved.

OCCASIONAL INFORMATION

Travelers will want access to information from sources they do not use on a day-to-day basis. In Boston, one might want to find out about restaurants and in Chicago, find out what shows are in town. Another might want to access a list of flights from one city to another within a specified period of time and then find out if a certain hotel chain has a hotel located near the destination.

Users can arrange for information they want and need each day in advance. They might even want their handheld mobile computer to access the information for them at a given time each day, or to notify them when data is awaiting retrieval.

Occasional information, on the other hand, will be important to us at a specific time, and this information should be readily available even though the user has not been able to anticipate and pre-arrange access. Users should be willing to pay a premium for this "occasional" use information, but they should be able to access it on a demand basis without having to notify someone to make it available to them.

ADDRESS BOOKS

Systems either already on the market or about to enter the market include an address book in addition to a calendar, to-do list, and memo function. The address book should allow users to enter information about a person. This information can include company affiliation, title, maybe the address, at least one phone number, and sometimes an e-mail address.

In reality, this information usually includes both a business and home phone number, a fax number, a cellular number, maybe a paging access number, and one or more e-mail addresses. What is not provided for is an indication of the best way to get in touch with people, how often they check their e-mail, if they are traveling, or other information necessary to make contact efficiently. Users simply have to rely on contacting their voice or e-mail, or sending a fax. The sender has no idea how long it will be before the intended recipient sees the message or if he or she is in a position to respond to the communication.

As wireless mobile computing advances, users will need to have some way of knowing the best method to reach people in the fastest and least obtrusive manner. AT&T is presently offering 0–700 numbers, but unless the calls are diverted to the user's cellular phone or to another number, it will ring in the office, connect the caller to voice mail, and the message will have to wait for the recipient to retrieve it. What is needed is a way to put messages up on a system and have the intended recipient "find" the message no matter which service he or she happens to be using at any given point in time.

FINDING SOMEONE

If a user needs to get a message to someone, the choices are to leave a voice mail, send a fax, send an e-mail message, write a letter, or even send a Federal Express envelope. The fax, the letter, and the Federal Express letter all require that the person be at the location where the package was addressed or that someone is at that location to forward the package. Voice mail and e-mail can be accessed from almost anywhere, but how does a person know that a message is waiting and how urgent that message is?

DEFINING THE PROBLEM

Wireless mobile computing not only means access from anywhere at anytime for anything, it also means a change in methods of communication. As people become wireless mobile users, they will, by virtue of their newly-acquired untethered status, find that since they are "in touch" more, from more locations, it is expected that they are, in fact, "findable." Business procedures and levels of expectations will change due to this new-found "instant access."

Cellular phones and pagers have already made many users captive to this new short communications turn-around. Once people learn that someone is "wired" on the road and can receive messages in short order—that a person is mobile but accessible—they will be even less inclined to wait until he or she "surfaces" before expecting a response.

AN ASIDE

When users first join the ranks of the cellular "connected," when they clip on their first pager, the expectation is that they will be more accessible. It is, therefore, very important that accessibility be defined and that the level of response be consistent with expectations

If users set a level of expectation that from the time they are paged or called they will respond within minutes, they will be forced always to respond within that time frame. Increased communication means increased access, and that can mean less, not more, productive time for the wireless mobile user.

Many of those "going mobile" have succumbed to the temptation to be instantly accessible no matter where they are and no matter what they are doing. The way users manage their connectivity from the very beginning will determine if these new wireless devices will truly serve them or if they become slaves to their devices, to the detriment of both their ability to get things done and their outlook on being truly "connected."

BACK TO THE MATTER AT HAND

Currently, remote services can be accessed through RAM Mobile Data, ARDIS, and even by using a cellular phone modem. But this only serves to move the data from one location to another; the user must then figure out what needs to be done with it. Further, unless the exact e-mail address of a person or company is known, there is no way to "find" and include him or her in the information loop.

ENTER GENERAL MAGIC

General Magic had been one of the best-kept secrets in Silicon Valley from 1990 when it was founded until its January 6, 1994, announcement in conjunction with AT&T, one of its major investors. During this time, it moved forward with product development and forming industry partnerships. At the time of the announcement, General Magic's alliances included Apple, AT&T, Matsushita, Motorola, Philips, and Sony.

Two software products were announced at the AT&T PersonaLink Services press event in January 1994. The first was Telescript, a communication technology that enables people to send electronic agents into the worldwide grid of information highways. Telescript is part of the long-term vision of General Magic, but there was another piece to the announcement—Magic Cap. Magic Cap is a software environment, a platform for communicating applications, and a foundation for personal communicators. In short, it is an operating system for handheld portable computers.

Telescript is a platform-independent, user interface-independent communications technology, while Magic Cap can be considered *one* of the front-end or user interfaces that can be placed on top of Telescript.

MAGIC CAP

Magic Cap was developed to provide everything a person needs to interact through today's popular modes of communications—fax, public electronic mail services, and telephones. Integrated with its core communication capabilities, Magic Cap software includes features to help manage personal information: address cards that are automatically updated as the sender's information changes, a calendar that issues invitations to meetings, and a notebook that supports free-form and structured notes. Magic Cap's interface represents real-world objects on a screen, guiding people through its expandable range of capabilities (Figure 6-1).

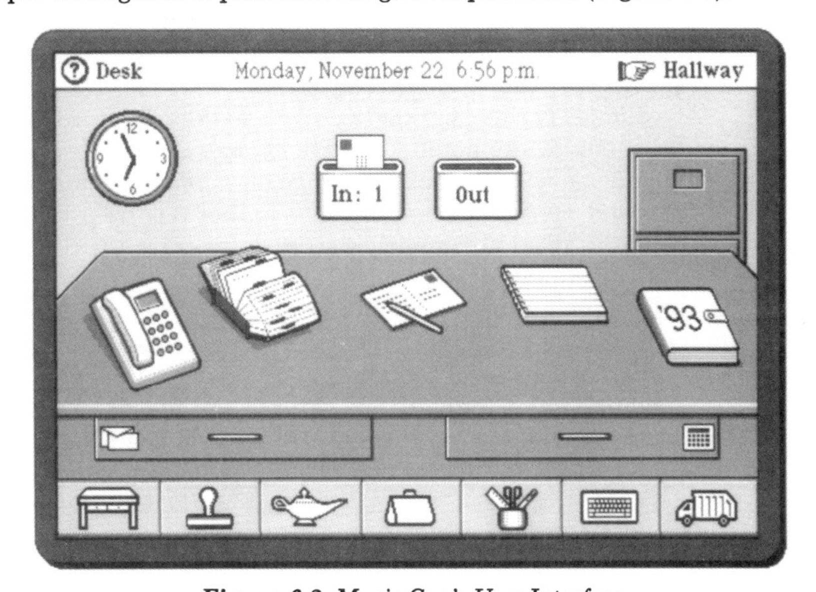

Figure 6-2. *Magic Cap's User Interface*

Magic Cap's major design consideration is that it relies on a pen or stylus input device for navigation and for writing notes in "ink." The program does *not* include any handwriting recognition. When an e-mail or other ASCII-based message needs to be composed, a user can make use of an on-screen keyboard or plug in an external device. General Magic believes that handwriting recognition has not yet matured to a point where it can be used in place of a keyboard. While "hooks" are included in Magic Cap for both handwriting and voice recognition, they are not required in order for a user to take full advantage of the operating system.

Magic Cap, as an operating system, will reside in 3 MB of ROM (Read-Only-Memory) in Personal Communicators, and it will be available in two software versions—one for Windows-based PCs, and one for Macintosh products. A user of a Magic Cap-based Personal Communicator can connect the device to a desktop machine and both will automatically be updated with the latest calendar, to-do, and address book information. Applications that are resident on the desktop system can be imported into Magic Cap and reconciled using a software product made by IntelliLink, Inc.

Magic Cap offers many unique features that will enable both computer-literate and non-experienced computer users to move around within the environment easily and smoothly. The real power of the Magic Cap operating system is its ability to make use of "agents" which are launched with Telescript (see below).

Agents that are either included with the device or acquired from an independent software developer can be set up to act as the user's surrogate. On Telescript-enabled networks, a user "sends" out an agent that has been designed to collect information or complete a specific task, or set of tasks. This concept is important when one connects to any network, but it is especially viable when connecting through a wireless link. By sending an agent out onto the network, the user does not have to stay connected while the information is being gathered. Instead, the agent is dispatched and, when finished, returns to the user's device with the task completed or the information requested.

An example of this might be an agent that has been set up to obtain plane tickets for a trip, make hotel reservations, list all the Mexican restaurants within walking distance of the hotel, arrange for a limousine for transportation from the airport, and provide the most recent weather report for the destination. Once dispatched, the agent accomplishes all such tasks by "visiting" the various places on the network where these services can be obtained. It returns to the user with all the arrangements made. On the day of the trip, another agent might be dispatched to the flight information service where it would sit and watch the airline schedule, updating the user if there were any delays in the departure time of the flight.

Magic Cap offers many other features and functions that it believes will help end users complete various tasks. Included in the basic operating system are a series of "rubber stamps" which can be used for many purposes. There is, for example, a stamp marked "urgent," another marked "confidential," and even stamps for holiday greetings. In addition to stamps, there are sounds for use on the local machine. By making use of the Telescript communications technology, such sounds and "stamps" can be sent along with a message. One stamp is the depiction of a man carrying a cake—the sound clip for the song "Happy Birthday" can be attached to the stamp. In this case, the stamp is animated and it walks across the screen while "Happy Birthday" is being played.

Magic Cap uses a highly sophisticated graphical user interface to provide a system that is easy to use. The view that will be used most often is that of a desk—complete with clock, telephone, rolodex, postcard (for messages), notepad, calendar, an inbox and outbox, and a file cabinet. Other views include a hallway down which the user can navigate to other rooms such as library, storage, and games. Leaving the "house," the user enters a street on which are buildings to which the user can via the network, "buildings" such as stores, banks, and libraries.

TELESCRIPT

While Magic Cap is the portion of the General Magic equation that the end user will see as the interface on devices from Motorola, Sony, Philips, and others, Telescript is the most important element of the announcement. Telescript is the underlying communications layer that must be installed in the handheld, on the network, and in computers tied to the network by service suppliers. It is Telescript that enables the users' agents to "wander" the network, searching for the information or services they have been created to find.

Telescript is an object-oriented remote programming language ("remote" meaning it must be installed at both ends). Its technology simplifies the task of distributed application development by abstracting the underlying complexity of communication—naming, addressing, routing, and security—into a high-level, platform-independent language. In Telescript technology, mobile agents GO to PLACES where they perform tasks on behalf of a user. Agents and places are completely programmable, but they are managed by security features such as *permits, authorities*, and *access controls*. Telescript technology is portable, allowing it to be deployed on any platform, over any transport mechanism, and through assorted media—wireline and wireless. Embedded in Telescript is the ability to handle different connection types, including text, graphics, animations, live video, and sounds.

THE IMPORTANCE OF GENERAL MAGIC

General Magic is one of the companies working on the links that will provide true transparent navigation capabilities to the end user. Both General Magic's Magic Cap and Microsoft's WinPad are based on the premise that the user's prime concern is to be able to obtain and work with information regardless of its form or where it resides.

It remains to be seen if a world with multiple standards, across multiple networks will be developed. General Magic will license Telescript to any network, information supplier, or hardware vendor that wants to make use of it. With the backing of AT&T, Motorola, Sony, and others, it certainly has a jump on some of the other vendors that will be competing to provide the necessary network intelligence to enable wireless connectivity to become easy to use.

It is uncertain whether General Magic will have to go head-to-head with Microsoft, Motorola, or even IBM in this endeavor, or if some or all of these competitors will align themselves and evolve a common, pervasive standard that will be usable across all networks and in every type of device. In any case, General Magic's products are one of the five enabling pieces needed for pervasive mobile computing use.

PIECE NUMBER TWO

In November 1992, when RAM Mobile Data announced the commercialization of its wireless data network, one of the companies that accompanied them at the announcement was a little start-up called RadioMail Corporation.

RadioMail is an important piece in the puzzle of information providers. RadioMail may not, in fact, be the only company that will provide the following services, but it is certainly the one that has a vision of what is needed. It has forged ahead to provide a working system model that is up and running today. The RadioMail system provides a number of the services that will ultimately be required, and it is available with both the ARDIS and RAM Mobile Data networks.

WHAT IS RADIOMAIL?

If the wired and wireless communications links are the pipes through which information is sent, RadioMail represents the reservoir and the switching station or the post office that controls the information put into the pipes.

One of the challenges of untethered communications is how to get to all the information that resides somewhere. Users might use MCI Mail and their own office e-mail system and want access to additional information services.

One way for communications providers to supply users with access to all this information is for each provider to contract with each and every type of information supplier that offers service and then find a way to connect users to each of these services on demand.

The RadioMail concept is that it will serve as the one system that will connect to all information suppliers. At present, it provides direct RadioMail-to-RadioMail connection to wireless users over RAM, ARDIS, and the Internet. It also provides connections to most of the other e-mail providers. If a user knows a person's address and mail system, the user can get messages to him or her over RadioMail.

Each mail service and information supplier uses different protocols and different formats for address and header information. A company such as RAM or ARDIS, in order to make available all the various services required, would have to provide translators between their own protocols and those of all the information and e-mail suppliers.

ENTER RADIOMAIL

Instead, with a single connection to RadioMail, all the information sent and received over the RAM or ARDIS network, in their specific formats, is received at RadioMail. The address of the intended recipient is determined, the service over which the information needs to be sent is determined, and the formatting and proper protocol conversions are applied. The message is then sent on its way.

RAM and ARDIS only need to establish one connection—to RadioMail. RadioMail, in turn, connects to MCI, AT&T, AppleLink, in-house e-mail systems, wire services for news, and all the other information suppliers.

End users are also enabled by RadioMail. Since RadioMail has (or will have) made all the wired and wireless connections, users can connect to RadioMail over any number of different wired and wireless options. The system will "recognize" the users and provide them with their own mail, and with other information to which they have subscribed or which is offered over the system.

RadioMail uses a store-and-forward method of messaging. Therefore, if a user connects to RAM in Los Angeles and sends half a dozen e-mail messages, when he or she steps off a plane hours later in New York and signs on to RadioMail again, replies and any new mail that is being stored is automatically sent to the handheld system.

All the features of RadioMail are not yet available. When one is mobile, one can send and receive intra-RadioMail messages using RAM or ARDIS and RadioMail. Users can choose to forward their MCI Mail to RadioMail so they can receive those messages while mobile. When they are in the office, they can reverse this process and forward their RadioMail to MCI Mail. More capabilities are forthcoming in short order. The folks at RadioMail understand the needs and wants of mobile people and have made great strides in providing all the services they will require.

RadioMail also provides NewsFactory and RadioFax. Soon it will be able to provide connections to office e-mail systems and to AT&T Mail, MCI Mail, hotel, restaurant, and airline reservation guides, stock market updates, and other services. Once a connection to RadioMail over any of the wireless carriers or over a wired link has been established, all such services will be instantly available, sent in a single format and protocol. If users want RadioMail to activate their pagers when it receives a message, it can provide this service. It will connect to the pager service and send a code, or even a digital message, notifying users of what is waiting for them in this information repository.

RadioMail, then, is piece number two in the puzzle that is fast becoming mobile information access.

MOVING INFORMATION

Piece number three, the ability to move information intelligently from one platform to another, is furnished by two different companies. One provides file-level capabilities and one builds on this, moving to record-level information exchange as reconciliation.

Traveling Software has been providing users with the ability to move program and information files from the desktop to portable computers for a number of years. The program, LapLink, is in use all over the world, and LapLink's multi-headed cable which enables connections between various desktop and portable computers is familiar to many who have been using portable computers.

Over time, LapLink has developed the means to provide users with the ability not only to move programs and files from one platform to another, but also to synchronize these files. A user can dial in, or direct-connect two computers via LapLink, ask the program to merge the contents of two sub-directories, and make them the same on both machines. Future enhancements will include capabilities well beyond those currently supported and will provide more levels of file synchronization than at present.

For users of wireless mobile computers, the ability to connect to their own computers and retrieve entire files is important. Making sure that the file on the mobile computer is in fact the latest version of the file is critical to the success of mobile computing. For example, if a presentation has been modified and updated, the user needs to know that his or her version is, in fact, the most recent one.

LapLink supplies this type of "smart" data transfer and is, therefore, half of our third puzzle piece.

In addition to Traveling Software's LapLink Professional product, several new companies have emerged with products that perform the same basic functions, but in slightly different ways.

One such company is Nomadic Systems, Inc., and its product is called "SmartSync." It is designed to enable a number of different Windows-based computers to contain identical data. The main difference between this product and the one offered by Traveling Software is that SmartSync does not send and receive all files residing on two computers. Rather, it "looks" at each file and sends *only* the changes to the other machine.

Yet another entry into this product group is the "AirCommunicator" by a company called AirSoft. It, too, is designed to provide file access and synchronization between computers. But this offering was designed from the beginning to make use of wireless packet systems. AirCommunicator permits a remote user to access all the files on a desktop or network system, and stores the directory of that system on the remote device. The user paradigm is to select the desired file by choosing it on the local system. The wireless connection retreives the file.

OPERATING SYSTEMS AND TRANSFERS

Some of the new mobile computing operating systems will offer transfer capabilities built directly into the operating system. Microsoft's "WinPad," and General Magic's Magic Cap are two examples. Further, when coupled with the capabilities provided by the company discussed below, end users will have a viable, robust method of not only moving files from one platform to another, but also of ensuring that all the information is up-to-date, and that there are no conflicts.

The second level of file and data movement is provided by a company called IntelliLink. IntelliLink has gained a reputation within the handheld computer market for understanding that it is often important not only to be able to send and receive entire files, but to work with information within a file.

Its present product offering permits users to connect any one of a number of different handheld computers and electronic organizers to a desktop system and move their calendar, to-do list, phone book, database, and memos between the two platforms.

The program does not stop there, however. In moving a calendar from any one of a large number of desktop Personal Information Management (PIM) applications (Lotus Organizer, Polaris PackRat, Borland SideKick, Arabesque ECCO, and more) to any number of handhelds including the Casio Boss family, Sharp Wizard, HP 95LX, 100LX, and more, it "looks" at each entry and combines the two files into one. If a user has entered a new appointment on his or her HP 100LX and his or her secretary has entered a new one on the desktop for the same time and date, the IntelliLink program will "find" this conflict and present it on the screen so that the user is aware of the conflict and can take action to resolve the problem.

Likewise, if an address book is moved from one platform to another, the user can be assured of always having the latest version. New entries added on both the desktop and the handheld program will be merged. Field mapping is also provided. When a user opens IntelliLink and chooses the desktop PIM or database and the handheld device that is being used, if the handheld device includes a PIM in ROM, the IntelliLink program will permit the "mapping" of each data field to the appropriate field in the other device.

If the desktop has a database that contains many more fields than are needed on the handheld, the user can map only the ones necessary, leaving the others intact in the desktop database.

INTELLILINK FUTURE

As the world of mobile computing advances, the ability not only to move data, but to reconcile it, will become more and more important. If a user is sent a new appointment over a pager (which is connected to a handheld or embedded into it), IntelliLink will enable it to be analyzed and sent directly to the calendar. At the same time, it will check to see that no conflict exists.

IntelliLink is also useful for its ability to move files from one program to another and from one organizer to another. When switching from PackRat to Lotus Organizer, for example, IntelliLink can transfer all the data. Not only is this done quickly, but since IntelliLink "knows" both systems, it is able to convert the information into the proper new format and place each piece of data into the proper fields in Organizer.

Likewise, when moving a database from one handheld to another, IntelliLink permits moving the data easily and without having to export from one file structure and import into another.

The "smarts" in the IntelliLink package will begin showing up in more and more products and, in conjunction with Traveling Software's LapLink and the other two pieces mentioned above, it will go a long way toward providing seamless information exchange between many different platforms.

THE FOURTH PIECE

Another piece of the picture is the ability to "find" people no matter where they are and no matter what system they use for mobile computing access. In its simplest form, it might be referred to as an e-mail equivalent of the phone company's "one-person, one-number" concept.

This identification system, promised soon, will require pieces of hardware and software to be installed in all the major components of the networks. A portion would be embedded in the mobile computing device, a portion in the hosts or desktop systems that will be communicating with the mobile computers, and a portion in the networks that will be providing the services.

Part of this new technology will allow existing desktop or system-related applications in a mobile environment to be enabled without the need to rewrite them. The most important part, however, will be a common user Identification number (ID). This ID will permit enabled users to "find" any other enabled users no matter which of the many communications networks they are using, and even if multiple networks are used. In reverse, it will enable end users to enter wired and wireless systems from any point—through any of the available services—and to be "found" by the system. This will permit users to receive information that is being stored while waiting for them to access the network.

LIKE A CELLULAR SYSTEM

The new technology that is needed might be likened to today's capabilities on the cellular network. A user who leaves the home service area is still able to make and receive calls in other areas. At the end of the month, one bill is received from the home area cellular provider. The other providers that have been used "know" where the user belongs and have forwarded the charges incurred to the "home" provider for inclusion in the normal bill.

Many cellular systems (but not all) are capable of locating users and, therefore, someone calling on a normal cellular number can be automatically switched to the area in which the user is located. The "enabler" in this type of cellular technology is the identifying cellular phone numbers. The cellular carriers' switching computers are able to "talk" to other carriers' computers to track us down.

These principles will soon become a model for the same type of inter-network communications in both the data and mobile computing arenas. When the system concept for such a one-number, one-person data system and software-enabling technology is implemented, the goal of go-anywhere, do-anything computing will be much closer to being achieved.

THE FINAL PIECE

Wireless data transmission methods will not be widely accepted until a high level of data security can be assured. On January 12, 1994, National Semiconductor (NSC) announced what it hopes will become a widely accepted and pervasive method of providing security for data. It will not protect those who discuss sensitive information over a cellular voice phone, but it will certainly handle the data portion of the equation.

There is a growing need to provide security across both wired and wireless networks. NSC entered the fray with a new technology called "iPower" that delivers the highest level of personal, portable data security commercially available, even across unsecured networks, at a low cost.

iPower technology uses an impenetrable hardware-based access card known as a "token" for securing and processing secret keys and data. Tokens provide a higher level of security than previously possible with software-only or network-based security schemes.

iPower technology secures users rather than networks. It can be implemented on any type of network, whether wired or wireless. The first implementation is housed in a PCMCIA Type II card that is assigned to a specific user. To activate the card, the user plugs it into a PCMCIA card reader and enters a validation code. After the card and code are acknowledged, the system enables the user to send and receive information that is secure, across unsecured as well as secured networks.

The iPower system integrates advances in three critical technologies: semiconductor, networking, and encryption. It has been designed to be fully compliant, and is built on industry standards such as the PCMCIA Type II specification and DEC, RRS, DESX, RC2, RC4, PKCS, CCITT, X.500, and X.509 encryption and transmission standards.

Since the iPower system is designed to be used in conjunction with PCMCIA slots, NSC announced a desktop PCMCIA reader/writer device for those who do not yet have PCMCIA capability. The system itself is housed in a Type II card and contains a high performance, 32-bit microcomputer, several embedded powerful cryptography engines, and long term non-volatile data/algorithm memory.

This combination enables the system to provide positive identification and authentication (certificates) through public key crypto technology, message privacy through block-encryption, secret key exchange, message verification, and non-repudiation (digital signatures). It also provides for secure storage of secret keys, algorithms, and personal and transaction data.

The National Security Agency is mandating personal, portable tokens for future U.S. government agency access to the new data highways. The combination of data security provided by the NSC system meets or exceeds all present and future government specifications, and yet it is easy to use and totally secure. These new PCMCIA-compatible tokens are called "Tessera" cards, after the token ancient Romans used as a ticket, tally, voucher, and means of identification.

iPower cards contain a special high-tech Security Processing Unit (SPU), and their circuitry is unbreakable by electrical or physical tampering. Attempting to reverse-engineer the card will result in the SPU chip becoming blank—it will not give up its secrets to anyone.

The idea is that users will carry these cards as their tokens, and access will be granted only when the proper Personal Identification Number (PIN) is used in conjunction with a card. If the card is lost, it cannot be used by another person, and it can be replaced only by the certification agency that issued it.

Using a series of encryption and public and private key systems embedded into the system, the card provides positive identification of the user; reliable system authentication of the user; hardware encryption privacy through support for DES, RSA, and other encryption algorithms; positive verification of messages to prevent alteration or transmission errors; secure authorization capability, including digital signatures for non-repudiation; and on-board transaction recording to improve security and enable off-line transactions and metering.

Private and public keys are used to send secured data that can only be retrieved by the addressee. Each person is given a *public* key. This key is like a phone number, and it is listed in standard network directories. When sending an encrypted message, the intended recipient's public key is used.

On the receiving end, a message can *only* be un-encrypted by use of a person's *private* key, a key that is not published and not known to anyone. (In NSC's iPower system, it is not even known to the end user or the vendor, since it is generated by the card when it is first installed.)

The DES standard is another form of encryption that is faster than the use of private and public keys, but it requires an encode-decode key to be passed between the two parties wishing to share data. While it is faster, anyone posessing the key can decode the message.

The iPower system uses both of these methods as well as a third method known as "hashing" to provide for totally secure communications while taking advantage of the speed of the DES system. With hashing, a user looks up the public key of the person to whom the message is to be sent. A random number generated by the program is then encrypted and sent to the user. This number is used as a key for a DES session. At the end of the session, a hash of a specific length that must match the hash on the receiving end is sent. If all conditions are met, and the data is sent, it can be verified that it was sent by the person who said he or she sent it, that it was received by the person for whom it was intended, and that no one else had access to it.

This system is so secure that it can be used for monetary transactions, legal documents, and other transactions that require absolute verification of transmission and reception and need to be handled in a secure way.

PUTTING THE PIECES TOGETHER

Each one of these five pieces is provided by a different company or group of companies. The real challenge will be how to package them into an efficient, easy-to-use wireless mobile computing platform. Much of the work on this integration is in process.

The first products to take advantage of these technologies will certainly not turn out to be the do-all, end-all systems that will evolve over time. They will be the first of an entirely new generation of products that will provide another step toward realizing the promise of mobile computing.

All these advances are coming, and it is especially encouraging that most of these companies are talking to each other and realizing that a single company or organization cannot possibly provide all the necessary pieces and parts.

Realization of the wireless dream is coming closer. But much remains to be done. These five pieces of the puzzle bring the goal considerably closer. The visions of many different players and companies will need to come together into a series of common pieces and parts in order to to the rest of the way in achieving the goal.

7

Wireless Services, Selecting Systems

Having reviewed the basics of wireless communications, the types of systems, hardware challenges, and complementary technologies that will maximize the wireless computing exierience, it is time to focus on the selection of systems.

If a user needs or desires to implement wireless data communications, there are a number of ways in which this can be done easily at a reasonable cost. A sales force can be wireless-enabled, as can service representatives, executives, and any other segment of a company that has the potential to become more productive as a result of being untethered.

PRE-PLANNING

As with every aspect of systems implementation, it is important to spend time and effort at the onset analyzing needs, comparing them to what is available today, and determining cost. The chosen strategy should be implemented in phases. Most importantly, do not rush the process.

First, users must define why they are interested in wireless data access. Do they believe such implementation will save money? Are they looking for a way to improve sales or service responses? Do they need access to key executives no matter where they are or what they are doing? Or are they simply curious about the potential benefits of wireless data access and want to implement a pilot program to investigate the possibilities?

Each is a valid reason for exploring wireless options—even to satisfy a curiosity about the technologies. Before proceeding to explore a move to wireless, several major points must be considered and, in some cases, defined:

- Determine who will use the systems, what they will be used for, and who will control them.
- Do not assume that each requirement for mobile communications needs to be fulfilled by a single system. It may be more beneficial to equip some people with cellular phones, some with pagers, wireless e-mail, or direct LAN connections. Some may be better off with a pocket full of quarters for pay phones.
- Consider the type of information that will be sent and received via wireless. If long, detailed service instructions, or sales order-entry from the field are needed, a different set of options will be necessary than for the user who needs to send and receive short messages or e-mail-type communications.
- Give careful consideration to the physical locations involved in the data transfer. If the users need to send and receive data only when they return to a vehicle (as with a delivery service) the systems requirements are different from those of sending data to and from a service technician standing in front of a copier in the basement of a building.
- Consider data transfer logistics. Are person-to-person communications necessary throughout the coverage area, or will the traffic always be from individuals back to a main location?
- In these early stages of wireless adoption, companies should not make such a large investment in any given technology that they cannot afford to modify or change direction over time—especially over the next four to five years. Today's technology is changing and improving rapidly.
- If investing in wireless technologies today, companies should make sure they can cost-justify the implementation on a short term basis or as a learning experience.

PLANNING

Once the decision to try wireless data communications has been made, the next step is to determine the type or types of service will meet the company's needs and then the suppliers of such services must be found.

The following explanations of each system available today should help in determining the appropriate services for specific needs. Remember, however, that implementing a wireless strategy does not necessarily mean implementing a single solution for all mobile requirements.

In this chapter, various types of wireless data systems will be discussed. Each type can provide a solution to some wireless needs, and each has an associated cost.

Data Security

Besides cost, security issues must be addressed. While these systems are not as vulnerable to eavesdroppers as cellular voice communications, security should be considered. To determine how much time, effort, and money are required for security, the following questions should be considered:

- Will the information that is being sent and received be of value to anyone else? If so, how valuable?
- Could such data, in the hands of others, cause severe business losses or embarrassment?

If the answer is "no" to these questions, the user needs only to be concerned with "normal" data security measures. If the answer to either question is "yes," however, the user will need to be careful in the selection of service and the type of data encryption employed (see Chapter 6).

Many corporate executives think nothing of picking up a cellular phone, or an Airphone, and talking about sensitive company matters. Calls can be intercepted on a standard wired telephone circuit, but not as easily as over the air. Most cellular phone users have experienced being in the middle of a conversation over a cellular phone and losing contact with the person to whom they were talking, or finding themselves listening in on another party's conversation.

If the information overheard is sensitive, and the subject of the conversation can be identified, anyone hearing it could capitalize on it.

The best rule to follow in the area of data security is to assume that someone is able to retrieve the data being transmitted, and act upon it. One final point here—if someone *really* wants access to certain data or to listen to phone conversations, he or she can. Those who make a living by stealing information are very good at what they do. Most security measures will keep out the casual listener, and maybe even the determined eavesdropper. But no level of security available today will keep out a trained, well-equipped professional.

DOWN TO THE CHOICES

Today's choices for wireless data transmission include:

- Using an existing two-way radio system and adding data.
- Adding data to an existing SMR (shared) radio system.
- Using digital pagers.
- Using messaging pagers.
- Sending data over analog cellular systems.
- Contracting with RAM Mobile Data.
- Contracting with ARDIS.
- Starting in 1994, making use of digital data over cellular systems as CDPD is deployed.
- Starting in1995, making use of Mtel's Nationwide Wireless Network (NWN).
- Starting in 1995, implementing Metricom's system.
- Beyond 1995, using Personal Communications Services (PCS).

Spectrum allocations for these services are as follows (Figure 7-1):

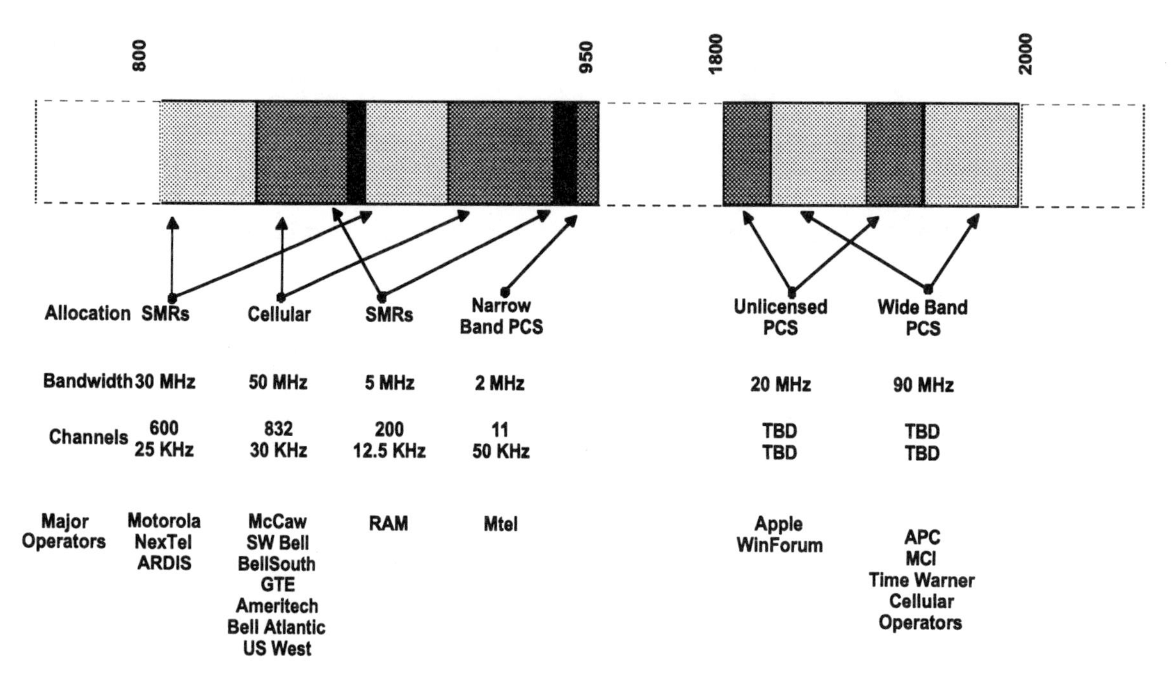

Note: Frequencies are in MegaHertz

Figure 7-1. *Wireless Data Spectrum*

Conventional two-way radio—base to mobile and mobile to base

- Low Band: 30–50 MHz, allocated in segments for all services.
- VHF (High Band): 150–174 MHz, allocated in segments for all services.
- UHF: 450–470 MHz, allocated in segments for all services. (Most two-way radio systems in this band use repeaters to extend the range of car-to-car communications.)
- UHF-T: 470–512-MHz repeater systems for two-way radio. (In metropolitan areas, two-way radio users must share these channels with UHF-TV channels 14 through 20—hence the name UHF-T, "T" for television.)
- 800-MHz Band: 800–940 MHz. This includes trunked two-way radio systems, cellular phone systems, paging (including nationwide paging channels), an unlicensed portion (902–928 MHz), and some point-to-point direct communications.

Each of these systems offers both advantages and disadvantages. And each is designed to provide coverage over different areas. The following discussion will take these options in order, interjecting geographic considerations into the choices.

LOCAL COVERAGE TO VEHICLES

Providing wireless data transmission capability to vehicles is probably the easiest to achieve. A two-way radio is used, with the addition of a data-modem in a vehicle. Complete mobile data terminals are also available for such applications.

Range calculations are also fairly straightforward. Generally, data streams will provide coverage somewhat further than normal voice communications. With acceptable voice coverage, chances are good that data coverage to vehicles will be equal or better.

The cost per vehicle will depend on the type of modems and terminals employed, and can run anywhere from several hundred dollars per unit to several thousand. The base station connection—to connect the radio system to the data system—will cost several thousand dollars. The result will be a dedicated data system integrated on top of your voice system.

Cautions

If there is a great deal of voice traffic on a system, the possibility exists that voice and data transmissions will interfere with each other.

Further, if a "simplex," or one-channel system is being used, the actual rate of data transmission compilations will be much slower than the speed of the modem. In a single-channel system, only one transmitter can be on at any one time. Thus, a packet of data is sent by one transmitter, and an acknowledgment must be sent back on the same channel before the next packet can be sent.

In a duplex configuration (two channels), the acknowledgment bursts can be sent almost simultaneously with the receipt of the incoming packet, thereby speeding up the entire process.

Advantages

The advantages of this type of system are:

- The user owns the system, and there are no recurring charges for data transmission.
- The system is not shared with others—unless a two-way radio channel is overcrowded with other users. It should be able to handle the addition of data without undue interference.
- This type of system is "controllable." That is, data transmission and collection activities can be controlled—just as a voice system is regulated by a dispatcher.
- Buying the equipment may be as expensive as some of the other options, but there are no recurring costs.

Once a system has been installed, or as it is being installed, some interesting mobile options can be used. For example, all the vehicles of a fleet could be wired in such a way that if any were stolen, the dispatcher could remotely turn off the ignition and blow the horn; a vehicle's vital signs, such as water temperature, oil, and water, could be measured and any conditions that exceed normal limits could be reported to the dispatcher.

In several bus data systems, dispatchers receiving alarm information have been able to notify the drivers of problems *before* they were aware of them, saving the bus engine from being damaged.

Global Positioning Systems (GPS) are also available. Not only can data be sent and received, vehicles can also be pinpointed to within about 30 meters of their exact locations. This feature would be particularly helpful to a pick-up and delivery service. Think of the advantages of being able to pinpoint the locations of vehicles, identifying the vehicle closest to a given call and its status, and being able to re-route it to make the additional pick-up.

Cautions

Before *assuming* a two-way radio channel can be used for a combination of voice and data services, users should check with their local two-way radio sales and service organization. Some radio channels can be used for both, some cannot legally be shared.

Summary of Two-Way

This type of data system has been in operation for a long time now (since the late 1960s and early 1970s), but the technologies have been improved over time. Such a system makes sense for delivery or service operations that are confined to a specific geographic area, and companies that already have two-way voice radio systems in operation.

SPECIALIZED MOBILE RADIO

Specialized Mobile Radio systems (SMRs) became popular for companies with two-way radio requirements in the early 1980s. Unlike standard shared radio systems—in which a number of different users have to share the same frequency—SMR systems are based not only on multiple customers with their own users, but also on multiple radio channels. Thus, more than one company can be operating on the system at once. The basics of the system are that each mobile unit "listens" to a control or command channel and, once requested, the system selects a channel for use, changes the frequency of all the units in the selected group, and then switches the entire fleet or units involved in the conversation to one of the operating channels. After the communications session has been completed, the units return to the command channel.

The benefit of this type of system is that it appears as though users are operating on a private radio channel, without interference from others. In reality, they are part of a system that may have hundreds of users. While it is possible to get a "busy circuit" indication, it is not very likely, given the number of users, the number of channels, and the odds that all the users might want to talk at the same time.

Early SMR systems have given way to complex, powerfully controlled systems that include private direct-dial telephone interconnections, paging services, and, yes, data transmission.

The first SMR operators were licensed on a city or regional basis. As demand for service has increased, many providers have begun to band together to increase their coverage area and to provide "roaming service" for users from other areas who are traveling through.

Fleetcall (now NexTel) and several other organizations have begun either buying up or combining forces with SMRs in other cities, offering larger coverage areas and more services. The roots of these systems were the same as most two-way radio systems—dispatch communications. Since most of the lower radio channels are "full," or congested, users desiring dispatch service for their vehicles turned to the SMR, or "trunked" radio channels.

There are many different types of trunked radio systems operating in this band—from complete city and county systems to systems designed to be used by many companies on a shared (and fee) basis. The ones with which we are primarily concerned are those that offer their services to a number of different companies.

Each SMR operator is licensed to operate from 5 to 20 or more channels in a given geographic area. One of the channels is used for coordination, while the others serve as communications channels. Each fleet is assigned a specific set of codes that identify the fleet, and define who is permitted to communicate within the fleet. When a dispatcher or mobile unit requires contact, the operator pushes the transmit button on the radio, sending a digital message to the control computer requesting a voice channel. The computer switches all the units within that fleet or subfleet over to that channel. Once the voice communications are complete, the units are commanded to return to the control channel and wait for the next request for communications.

Added Features

Not long after the implementation of SMR services in 1983, some SMR operators began adding capabilities to their systems. Today, there are systems over which is it possible to make and receive telephone calls, alert a specific mobile unit, and even to send and receive data.

As the popularity of cellular phones has increased, and the interest in mobile data has also increased, many SMR operators have begun offering both voice and data systems. At present, there are several SMR operators working on combining their systems to provide large regional systems with an eye on providing nationwide coverage over time. NexTel has been one of the major leaders in this area and has designed the interconnections necessary to unite a large number of SMR operations together into a much larger system.

Data via SMR Systems

Some SMR operators believe they will be able to build a nationwide system to compete with cellular phone services, nationwide data systems, and even the proposed PCS systems. However, since most of these systems were designed to provide coverage to mobile and portable units outside buildings, there is a question as to whether or not these systems will be able to provide the in-building coverage that will be required for full voice and data service to the next generation of handheld computers and telephones.

SMRs can provide data transmission capabilities to and from vehicles at reasonable prices with good coverage. By installing a data terminal in a vehicle and connecting the terminal to a computer, or having the data terminal act as a "mother ship" for small handheld portable data collection terminals, SMR services can provide reliable data transmission and reception. Further, since trunking systems assign a single channel to a user until the session is completed, they can often provide data transmissions that are more reliable and faster than those sent over analog cellular systems.

As with standard two-way radio data systems, there remain the problems implied by a lack of data standards, and the proprietary data formats of different vendors. Customers of most systems installed to date have become "captive" to a single vendor's system and are at the mercy of that vendor when it comes to system enhancements and improvements.

Selecting an SMR System

If a user's requirements are for data transmission to and from vehicles, *and* the use of voice for dispatching or other functions, discussing system options with SMR operators makes sense. There may, in fact, be cases where a data-only SMR system might be more economical in the short run, since SMR fees are generally based on the number of vehicles in a fleet, rather than the amount of airtime used. Other metropolitan and wide-area network data systems usually charge on a per-message basis, or on the number of packets sent and received.

SMR services, at present, may offer a real advantage in cost because of these differences. However, the basic equipment costs, installation, and on-going service could eat up any potential savings—which is why this type of service is recommended only if the requirements are for a combination of voice and data, and only if there is a finite regional area where coverage is required.

How to Find SMR Operators

Most SMR operators are two-way radio sales and service organizations, or at least are affiliated with local two-way radio sales and service organizations and can be found in the yellow pages. Unlike cellular, paging, and computer stores, most of these organizations have been providing two-way radio sales and service within a community for a number of years. They are also probably well versed in standard two-way and repeater operations, and know a lot about the finer points of two-way voice systems.

They may not, however, understand data transmission over their systems. Unless the operator already provides data transmission over its system, it is probably wise to find an operator experienced in this area. In locations where NexTel has an operator, or is working with an operator, the best bet would probably be to contact NexTel and use its system. If not, several other companies also provide data equipment for SMR operators.

There are several companies, RacoTek and QualComm, among them, that specialize in providing voice, data, and even vehicle location services over SMR sytems in the United States. Some of these services are already available on a nationwide basis, while others will become available over the next few years.

Recap

Thus far, data transmission using existing two-way radio systems has been covered. Before moving into paging, one-way and two-way data networks, and cellular systems, the basic points should be listed:

- Using existing two-way radio voice systems and adding data makes sense when sending and receiving data to and from vehicles is necessary. Most two-way systems, including SMR systems, have not been designed for in-building coverage and may not work well for such applications.
- Data which only needs to be sent and received from a vehicle does not mean that the user only collects or makes use of data while in the vehicle. Federal Express and UPS collect data in the field. Fed Ex has its own private mobile network while UPS utilizes cellular. In both cases, the data is sent to headquarters when the driver returns to the vehicle and inserts the data collection device into the data transfer unit.
- If a voice channel is heavily used, or if there are a lot of interfering signals present on the frequency, it is not a good idea to try to send data over the channel.
- If an existing radio system is a simplex system (base station and mobiles transmit and receive on the same channel) it is not a good idea to try to use the channel for both voice and data. The reason is simple. As an illustration, picture one of the mobile units ten miles south of the base station trying to send data to the base station. Another mobile unit is ten miles north of the base station and probably cannot hear the southern unit's transmission. Hearing nothing, the northern unit tries to call in using voice—interfering with the data transmission in progress. Numerous re-tries could be required before the transmission is successful and these re-tries would tie up the frequency.
- Data is not permitted on all two-way radio frequencies. Check with the FCC or your local radio service shop to determine if it is legal on your channel.
- Whenever possible, try to find another radio channel that can be dedicated to data-only transmissions. Many police and fire departments using mobile data systems employ a data-only channel because of the high level of voice traffic on their dispatch channels.

- Before investing heavily in adding data communications to an existing two-way radio system, the user should make sure it will provide the level of service required. If the data is time-critical (meaning it is needed within seconds or minutes), a dedicated radio channel should be used and this channel should be separate from that used for voice traffic.
- Make sure any data-related applications are written to provide error checking and that they send and receive *only* the data needed. There is no need to send a complete form back to the base station—only the data that has been collected in the field needs to be sent. Sending the entire form takes more air time and is wasteful.
- Data over an existing voice system can be more cost-effective than almost any other means if a company already owns its own two-way radio system. It will only need to add data capability.
- Most data systems designed for existing two-way radio systems are proprietary—meaning the user will be locked into a single vendor for its data requirements. The user should be sure to obtain alternate proposals and that the system performance carries some type of guarantee.

TWO-WAY DATA-ONLY

Two services that are fully operational are the ARDIS system and that of RAM Mobile Data. The services in the process of being implemented are CDPD (the transmission of digital data using the cellular network) and NWN by Mtel. The most recent entry into the data-only wireless world is by Metricom, a company that has been providing wireless data transmission equipment in the unlicensed bands for more than seven years.

These are not the only services that will emerge over time. However, they are the ones that are already on-line or that will come on-line shortly. Systems still on the drawing board depend on new FCC rules for PCS, or on obtaining a license prior to construction of their backbone systems.

Others are actively engaged in designing systems for implementation. Several companies are running pilot and experimental systems in some areas of the country. Some such systems combine digital voice and data modes. Companies such as MCI are lining up partners for their own entry into the wireless data network business, as well. Sprint has aligned itself with Iridium, the Motorola world-wide satellite system that it hopes to have operational in 1996.

It appears, at the moment, that as the world progresses toward go-anywhere, access-anywhere, wireless computing, there will be at least a dozen more ways to wirelessly connect devices. The discussion here is limited to what is practical today and what will become available in the near future.

Initial Focus

ARDIS and RAM Mobile Data are available to end users today, and each provides wireless data service over a wide area of the United States. There are many debates about which system is best, which offers better service to end users, and which is capable of handling more users than the other.

For the moment, the issue of system loading is a moot point. ARDIS claims to have around 40,000 users and the RAM system less than 10,000—but growing rapidly. Neither is in any immediate danger of being overloaded, unless users flock to these services in droves overnight. Both have plans to expand service as demand increases.

The Differences

Before describing each system, it is helpful to differentiate between the two systems. The ARDIS system was initiated in the 1980s as a joint venture between Motorola and IBM to provide wireless data access to IBM's field service technicians. It was licensed on a single nationwide 900-MHz channel, but was operated on a regional basis. IBM service representatives did not need nationwide coverage—only the ability to send and receive data to their regional dispatch and support centers.

After the decision was made to broaden the scope of ARDIS and to make it available to additional subscribers, it continued its regional approach. Most of its customers still met the IBM field service model—the prime use was to dispatch service and sales forces. ARDIS provided wireless connections for rental car companies (such as Avis) and other customers who, while having a nationwide need for access to wireless, did not have a requirement for communications capabilities between wireless devices anywhere in the nation.

RAM, on the other hand, was designed from the beginning as a nationwide system. Its infrastructure was built to include the capability for seamless roaming. RAM, therefore, attracted users who need to communicate between wireless devices on a nationwide basis.

Both systems are presently undergoing changes. ARDIS has added nationwide roaming. RAM is targeting not only individual customers who need nationwide access, but regional customers who need one common nationwide system, but not necessarily nationwide inter-communications.

ARDIS was designed to provide service to and from a person in the field with his or her own computing environment. If an individual subscribed to ARDIS, there would have been no one to communicate with unless he installed a connection to his or her office computing system. RAM's service operates in much the same way.

Enter RadioMail

RAM, and then ARDIS, discovered that the type of messaging clearing house service provided by RadioMail gave them access to individuals—not just companies. With the RadioMail service, end users can subscribe to a combination of RAM or ARDIS and RadioMail to obtain wireless access to most public e-mail systems, as well as news and other such information.

Systems are in place and communications hardware—while not as small and light as we would like—is available at sub-$1,000 price points. Service prices are in the under-$100 per month range, and the networks are ready. What is needed now is a way to attract many more customers who want to gain from the advantages of wireless connectivity. RAM and ARDIS are both on the right track. Both are engaged in marketing activities to persuade new users to cut their ties and try wireless. Time is a critical factor in this quest. Both RAM and ARDIS need to add to their wireless user populations before other services that will compete for the same group of early adopters come on-line. How each goes about attracting users is as important as the basic differences between their networks.

Wireless Spectrum

Both RAM and ARDIS operate in a portion of the spectrum that requires them to hold licenses issued by the FCC. The ARDIS system operates in a portion of the spectrum located just below the cellular telephone channels. Each of its channels has a bandwidth of 25 KHz. (As a rule of thumb, the wider the channel, the higher the possible data speeds.) The RAM system operates in a portion of the spectrum just above the cellular telephone channels and its channel bandwidth is 12.5 KHz (half of the ARDIS per-channel bandwidth).

The ARDIS system is built on a single nationwide channel augmented by additional channels in areas where usage is higher than a single channel can handle. The RAM system, on the other hand, is licensed for a total of 30 channels per area. It has installed between 10 and 30 channels in each coverage area.

The differences in frequencies, bandwidth, and the various channel-switching schemes make it almost impossible to build a single transmitter and receiver combination that would enable a user to operate on both the RAM and ARDIS systems. CDPD (Cellular Digital Packet Data), which will be implemented during 1994, will make use of existing cellular channels with yet another bandwidth and channel spacing. NWN will make use of yet another bandwidth. For these reasons, users must first determine which service they are going to use. **IT WILL NOT BE POSSIBLE TO MOVE FROM ONE TO ANOTHER WITH THE SAME EQUIPMENT.**

Users of cellular voice phones can change carriers within a cellular service area. (There are two cellular service providers in each coverage area.) The FCC requires all cellular phones to be capable of operating on all cellular channels. Moving from one service provider to another is simply a matter of obtaining a new phone number and having the phone programmed to change carriers. In the world of wireless data transmission, this option may never become available.

Since users will not be able to simply switch wireless communications carriers at their whim, it is important to understand the strengths and weaknesses of each of the systems, and to carefully choose the appropriate supplier.

ARDIS

Today, the ARDIS system provides coverage in more than 400 U.S. cities, accounting for 80 percent of the population and 90 percent of the business-related activity. When ARDIS enters into an agreement with a corporate end-user fleet, it provides a radio coverage contour and contractually guarantees that level of coverage. Further, because its system was designed from the ground up to provide this type of coverage for IBM field engineers who work on computers, copiers, or similar machinery, ARDIS claims it offers the best in-building coverage available in the areas it serves.

ARDIS coverage is accomplished by making use of more than 1,300 radio base stations controlled by 35 RF/NCPs (Radio Node Controllers), which, in turn, are controlled by 3 switch nodes. (A fourth will soon be brought on-line.) Each switch is located in a hardened telephone company site. The three sites are tied together using a series of dual-diversity routed circuits. Each switching node is comprised of a Tandem four-processor VLX, a two-processor Cyclone, and direct and dial-up switching units in a redundant configuration.

The switches are located in Chicago, Lexington (Kentucky), and Los Angeles. The LSAs (home of the RF/NCP units) are located in White Plains, New York, Washington, DC, Atlanta, Chicago, Dallas, and Los Angeles.

The system is interconnected by multiple-route T1 lines, and the circuits are constantly monitored by ARDIS and AT&T. ARDIS claims it maintains better than 99.97 percent availability of the network, and averages 4 seconds per message response time (round trip response time). At present, there are (according to ARDIS) 40,000+ subscriber units and 300 customer hosts on the system. Most of the hosts are direct-connected to the network using telephone company lines, but some have their own RF link—either as prime access or for back-up purposes.

Radio System Deployment

Again, much of the ARDIS system makes use of a single radio channel. In any given service area, there are multiple radios deployed to provide both in-building and on-street coverage. ARDIS refers to this single-channel deployment as its single-frequency layer scheme. Through re-use of the same channel, it can provide many simultaneous "data conversations" while providing excellent in-building coverage.

In an example furnished by ARDIS, it cites the use of three base stations that all carry the same information intended for a single subscriber unit. ARDIS calculates the probability of reception by the subscriber unit to be 50 percent from radio one, 50 percent from radio two, and 70 percent from radio three. Using these numbers, the total probability that the message will be received by the subscriber unit is 92.5 percent.

ARDIS claims that the base difference between its system and that of RAM is that a single-frequency system will provide better coverage over a given area than a multiple-frequency system that has been designed to provide for more capacity. In areas where ARDIS needs additional capacity, it has added channels. However, unlike the RAM system, such channels are used to provide different levels of coverage. In New York, for example, ARDIS uses a total of five different frequencies.

Two channels are designed to provide in-building coverage with 26 and 39 individual base stations, respectively. Another channel has 60 base stations and is designed to augment in-building coverage. Two more channels, using 10 base stations each, are designed to provide on-street coverage.

In Los Angeles, the ARDIS system uses two channels for in-building coverage, one with 23 base stations and the other with 16. On-street coverage is handled by a third channel with 34 base stations. Channel loading is carefully monitored in every city it covers. When the traffic load increases, channels are added, either to the in-building or on-street configurations.

Channels

ARDIS has a minimum of two channels licensed in each area covered. Six channels are presently available in New York and Los Angeles; five in Chicago, San Francisco, Philadelphia, Washington, DC, and Houston; four in Boston; and three in Detroit, Dallas, Miami, and Atlanta. While there are no "new channels" available, ARDIS can gain access to channels licensed to others within a given area by means of trading or otherwise providing incentive for a license-holder. (Licenses may not be "purchased" according to FCC rules, however, they may be "obtained" by other creative means.)

Data Speeds

The existing ARDIS network is based on a data rate of 4,800 bps (bits per second) and all the older subscriber units are locked onto the main channel with crystal-controlled units. However, all the new ARDIS equipment is capable of automatically switching between channels and will also be able to operate at the new high-speed rate of 19,200 bps.

The move from 4,800 bps to 19,200 bps represents an increase in throughput of between four and six times. This throughput improvement is accomplished not only by the increase in speed, but also because ARDIS makes use of a different signaling technique in the faster system, and several other differences. The result is that the packet transmission length is reduced from 23.3 milliseconds (MSEC) to 7.2 MSEC.

New subscriber equipment will enable users to take advantage of the new high-speed data system as well as to remain compatible with the older 4,800-bps system. ARDIS is implementing the higher speed mainly on the channels it is adding in the major-use cities. As a subscriber comes on-line in any given city, the subscriber unit is "told" by the system which channels are in use and what data rates are usable on which channels.

Subscriber Units

Existing subscriber units are limited to two frequencies at the 4,800-bps rate. New subscriber equipment is either 19,200 bps-capable only, or dual protocol, making use of all the available channels by switching between 4,800- and 19,200-bps modes.

The first subscriber units were integrated terminals and radio devices. Since they were designed for specific field use, they did not have to be compatible with other computing functions—nor were they required to run off-the-shelf software.

Today, ARDIS offers a radio and modem combination made by Motorola, called the "InfoTac." It can be connected to virtually any DOS or Macintosh computer, and it provides access over the ARDIS network to a company's own network, or, with a RadioMail connection, to the Internet and mail services such as MCI and AT&T Mail.

The InfoTac unit differs from the unit offered by Ericsson (which is also private-labeled by Intel) in that in addition to offering connection to a computer, it can hold a limited number of pre-programmed, "canned" messages that can be transmitted back through the system without having to connect the device to a computer. Also unlike the Ericsson unit, the InfoTac includes an LCD display window so that short, incoming messages can be read directly from the InfoTac.

When Motorola PCMCIA card radio modems become available, any portable computer with a Type II PCMCIA slot will have the capability to become an ARDIS subscriber unit.

Support Activities

One of the major areas in which ARDIS is trying to distinguish itself is in customer training and support. It maintains two "command and control" centers (one as a hot back-up of the other). Customers can call these centers twenty-four hours a day, seven days a week, and talk to a trained systems technician. ARDIS claims that 85 percent of all trouble calls it receives are resolved with a single call. The rest are resolved during a second call.

The control center also monitors the entire network. An operator sitting in the command center can check the status of any component of the system, and can "see" out over the network to any given radio base station. If a radio cabinet door is opened, or site access is attempted by anyone, the command center is immediately notified. Likewise, if a base station or any other component—including a telephone company circuit—develops a problem, the command center is notified and the proper repair organization is dispatched to correct the problem.

Seamless Roaming

ARDIS is in the process of implementing seamless roaming on its system. Previously, users had to arrange to receive service in cities you planned to visit. Now, such notification is automatic. When comparing ARDIS with RAM, two distinctions are that RAM is ahead of ARDIS in the area of seamless roaming because its system was designed to permit roaming from its inception. On the other hand, ARDIS has better coverage than RAM in more areas of the country.

ARDIS is now implementing seamless roaming and RAM is increasing its coverage. While there may be differences at the moment, both companies are making significant progress. Neither shortcoming should be considered a detriment for more than the short term.

Conclusions Regarding ARDIS

The ARDIS system was developed to provide wireless data connections to and from specific companies and their own fleets of mobile users. ARDIS has been very successful in providing this type of service. It realizes, however, that if it is to be a major player in the world of wireless data transmission, it will have to provide wireless connectivity to individuals, not only those who want to connect to their own company's computing assets, but also to connect to nationwide messaging and information services.

It is for this reason that ARDIS is implementing seamless roaming. It is also why it has teamed with RadioMail to provide service to individuals as well as companies. ARDIS is expected to become very aggressive in its marketing efforts, offering introductory wireless connectivity packages that will include an InfoTac and a RadioMail account over the ARDIS network. ARDIS is also expected to offer other types of turn-key solutions for those who want to experiment with wireless connectivity.

For those who want to enter the world of wireless, ARDIS is a viable, inexpensive way to find out if wireless technology is all it is cracked up to be.

RAM Mobile Data

RAM Mobile Data recently completed its system build-out for the top 100 U.S. metropolitan areas. RAM officially began offering service in the United States in December of 1992. Its system is based on the Mobitex network pioneered by Ericsson in cooperation with Sweden's National Communications Authority.

RAM completed the first phases of its system build-out in the United States in June of 1993. It serves the top 100 Metropolitan Statistical Areas (MSAs) with more than 800 base stations, covering 90 percent of the U.S. urban population as well as major airports and business corridors.

Mobitex systems are operating in the United States, Canada, United Kingdom, Sweden, Norway, and Finland, and are being installed in France, Holland, and several other countries. The RAM Mobile Data system in the United States presently serves somewhat less than 10,000 customers nationwide.

As mentioned, the RAM system is based on standard trunking techniques. This means that unlike the ARDIS system, RAM base stations are "smart" in that there is a "discussion" between the portable and the base station. Each base configuration has from 10 to 30 channels per location.

The system's architecture is driven from the National Control Center. RAM's system was designed from the ground up to provide both store-and-forward capability and seamless user roaming. The master switch controls the system that runs out over long-distance links to each MSA.

Each MSA is controlled by its own host computer and a series of local switches to the base stations. Because the system uses true trunking technologies, anytime subscribers are connected to the system, they are switched to one of the radio channels and packets of data are exchanged. Multiple users can maintain communications on a single radio channel. Thus, with a minimum of 10 channels per location, RAM has the needed capacity from the start. One of the major economic differences between ARDIS and RAM is that RAM is supporting a larger system capacity from the beginning, while ARDIS is adding capacity as it goes along.

The investment per city is initially higher for RAM. But ARDIS may have to scramble to "find" additional channels as they are needed for its major coverage areas. So far, this has not been a problem. With RAM already licensed and ARDIS having to beg and borrow existing channels, this could become an issue in the future.

RAM's math for its channel loading is that it has an average of 2.5 channels per installed base station. Each channel can handle an average of 500 users without system degradation. At present, neither system is threatened with overcapacity and both companies are committed to providing expansion in areas where service demand is high.

The other RAM feature that has helped push it into a strong competitive position with ARDIS is that its system was *designed* to handle roaming. The ARDIS system is adding roaming capabilities to its existing system.

Since all the RAM channels are identical in each area of the United States, a single wireless modem can make use of any and all channels in the system. In the case of ARDIS, users *might* be limited to the two common channels (using older equipment) and find it necessary to upgrade their units to make use of a variety of channels employed in various areas.

RAM, an "Open" Standard

When RAM first rolled out its system in October of 1992, it stated that it would "succeed in the wireless messaging market by providing networks optimized for message traffic and by helping developers provide wireless connectivity in their products and services."

RAM has been successful in this approach. It was the first of the major data networks to sign an agreement with RadioMail, as well as PSI—two companies that provide store-and-forward and messaging. Users have access to public e-mail systems, as opposed to captive corporate e-mail access.

Thus, while ARDIS has grown by promoting additional corporate access, RAM has been increasing its market size by enabling individuals to take part in the wireless revolution.

Until recently, this was a major distinction between the two networks. RAM was the enabler for individuals who wanted ties not only to their own corporate e-mail systems, but also to public networks. ARDIS concentrated on connecting users within companies to the specific company's computing resources.

Now ARDIS is offering RadioMail service. And RAM is working with key corporate accounts in an effort to increase its subscriber base more quickly. For all practical purposes, the two services are equal in providing access to public e-mail services. Both systems also charge approximately the same rates on a monthly basis (between $70 and $90, depending on the connection).

Pricing Models

In order to encourage new users to try wireless communications, both RAM and ARDIS have offered special flat-rate subscriptions to their services in the $70–80 per month range. Beyond this initial trial period pricing, only RAM has published a firm pricing schedule, based on three tiers of flat-rate service:

- Consumer Plan: $25 per month for up to 100 KB (kilobytes) of messages. Additional messages are $.25 per KB.
- Mobile Professional Plan: $75 per month for up to 400 KB of messages. Additional messages are $.20 per KB.
- Power User Plan: $135 per month for unlimited messaging.

Early experience indicates that the average size of an e-mail message (including network overhead) is about 1,000 characters long (or about 1 KB). Based on this average length, a "Consumer Level" user could send and receive 100 messages per month for the basic $25 monthly fee.

Always the Lowest Price

Users who sign up directly with RAM Mobile Data for a one-year period will have the best of both worlds. Each month, RAM will calculate the charges using all three plans and bill at the lowest possible rate, based on that month's volume.

A Word on Pricing

While RAM introduced the industry's first attempt at a pricing model, it will not be the last. It is important to understand that if a user is not subscribing to service directly from RAM, its monthly rates may be different.

For example, if a user signs up with RadioMail and chooses the RAM or ARDIS network (since RadioMail works on both), it will be billed by RadioMail and not by RAM or ARDIS. The bill will include actual airtime used or number of packets sent. It will also include the price of the RadioMail mailbox and any other services ordered from RadioMail. In this case, RadioMail buys airtime from RAM or ARDIS and resells it to the end user in conjunction with its own value-added services.

Key Differences

Presently, the major differences between the networks are that RAM's seamless roaming service is up and running as part of its basic system. ARDIS is still in the process of bringing its roaming system on-line. ARDIS appears to have slightly better coverage in areas of heavy corporate account concentration, but RAM's coverage is more than adequate for most computing activities.

While the population of users is greater on the ARDIS system, RAM appears to be on the brink of out-marketing ARDIS. ARDIS is beginning to offer RadioMail to its users, but RAM has made a deal with Intel that will provide almost instant access to its network for those who want to experience wireless messaging—without disturbing their corporation's computing environment.

This is the plan: Intel is marketing and private-labeling a RAM-compatible wireless modem made by Ericsson. This wireless modem is available in all retail outlets that sell Intel products—including its line of networking cards.

The potential user has to make only two choices under this purchasing model. The first decision concerns the mail connection software that will be used. Because of agreements with Lotus, Microsoft, and AT&T, the choices today are cc:Mail (Lotus), Microsoft Mail, or EasyLink (AT&T).

The only other choice that needs to be made is whether to run an X.25 telephone line from the server to the nearest RAM facility or to use a second wireless modem. The recommended choice is to purchase two wireless modems. Intel sells one that is configured to run from a standard AC power line and has a standard serial port connector for direct connection to your LAN. This configuration is easy to install and the Information Services (IS) department should not have to worry about making any changes to its system.

Once the software is installed on a mobile computer, and the wireless modem is connected by way of the serial port, the mobile user is ready to experience wireless e-mail. The total cost for this two-modem configuration, and the e-mail remote package, is under $2,000. A subscription to RAM on an unlimited monthly basis is less than $75 per month.

This configuration will not provide a user with access to anything other than his or her own server and mail system, but if the test proves successful, services such as RadioMail can be added easily enough with a little configuration change. RadioMail can provide the same access to a server and e-mail in either of the two ways mentioned above. It can also provide connections through the Internet to AT&T Mail, MCI Mail, and other public e-mail services as well as news, weather, and sports information on demand.

The advantage of adding a service such as RadioMail to a user's system is to provide access to other services and other people who may be on public e-mail systems. If a user's needs will be confined to his or her server access, a RadioMail connection will not be necessary. ARDIS can offer the same type of plug-and-play service, but, as yet, the InfoTac wireless modems are not available in retail stores in a shrink-wrapped configuration.

Marketing is Everything

Well, not really, but building a new market is part technology, part marketing, and part distribution. The Intel/RAM/Lotus/Microsoft agreements should help those who want to learn more about wireless e-mail to take the first plunge. Such a plunge does not have to be expensive, yet it will prove the value of wireless connectivity.

DATA OVER CELLULAR

There is one more type of existing wireless infrastructure that can be used for data transmission—the standard analog cellular phone system. Many cellular phone users have tried sending faxes and data via cellular phones. Some seem to think it is an acceptable method, but those who need to get information through every time, without frustration, find it is not as easy as it sounds.

Cellular and Wireline

Since a cellular phone works very much like a wired phone, except that the signal travels over radio waves, one would think a standard wireline modem could be connected to a cellular phone, a number could be dialed, and the connection could be made just as with a standard telephone.

It is, however, important to understand the differences. Once a wired system is used to access a remote data point, the line (the connection) belongs to the user for the duration of the call. There may be some noise on the line, but generally lines are clean, and once the various telephone switches have been connected, users usually stay connected.

This is not the case in the world of cellular transmission. First, there is the matter of radio channels versus wired connections. The cellular system uses 832 different radio channels (half assigned to each of the two carriers in a given area), and each system is made up of a number of "cell sites" that use some of the channels.

Users connect to the cellular system from the cellular phone to the cellular site via radio. From the specific site, the call is routed to a master system switch over either radio (microwave) or wired connections where it is then handed off to the standard telephone system.

The portion of the connection from the cellular master system to the standard phone system works like a standard phone system. Inside the cell system, however, the routing can be changed numerous times during a call. If a user moves even slightly while using a cellular phone or drives a short distance, the call may be shifted to a different channel in a different cell. Even sitting still does not guarantee that the user will not be subjected to cell switching. This can occur because the system is constantly monitoring the quality of the circuit. If the circuit starts getting noisy, the system can make a determination to switch the call. Another reason for a switch is that someone else entered the same cell and was switched into this cell. The existing user may be switched to yet another cell to adjust the load.

Switching times are not really discernible in a voice call, but they can run as long as 300 milliseconds—enough to put a glitch in data sent at 9,600 baud or even 2,400 baud. For this reason, a series of special modems that use error-correction and data-compression schemes has been developed. The most common of these protocols is the Microcom MNP error-correction protocol. MNP version 5.0 is the currently accepted standard. Most modems designed for cellular data over existing analog phones are making use of MNP 5.0, some are becoming available for

MNP 10, and AT&T is close to releasing its own form of error-correction and data-compression formats that it hopes to make standard.

Not only are there likely to be data interruptions during a cellular transmission, the end user remains connected to the host computer in a session mode—that is, the connection is maintained during the entire data exchange sequence. Since each character is sent over the system as it is typed and echoed back for confirmation, the user sees all errors as they occur. The frustration level can be high. Further, the cost of using cellular services at $.30 to $.40 per minute makes a standard session connection of five minutes at least a $1.50 call. When using standard wired circuits, the same call might be local or toll-free.

Sending faxes over cellular circuits seems to work acceptably. Once again, however, a user must be concerned with the direct costs involved. While sending a single-page fax to a single location is probably not any more expensive than sending it to a hotel that charges a per-page fee, sending and receiving many faxes can drive up the monthly cost of a cellular phone.

CELLULAR DIGITAL PACKET DATA

Often touted by major computer companies as being the "real" beginning of wireless data communications, CDPD is a technology that overlays the existing analog cellular system. The idea is simple. It will make use of the millions of dollars' worth of equipment already installed to provide wireless phone service throughout the major population areas of the United States. Adding data capability will increase the revenue stream for cellular systems providers, and will provide customers with the ability to use combination devices that support both voice and data.

Today, CDPD is a technology, a way to move data over wireless networks; it is not a wireless data network. This is an important distinction. Another important distinction is that each of the other systems is being driven by a single company. CDPD technology is being implemented by a number of different companies.

The Concept

The CDPD concept was developed by seven leading suppliers of cellular telephone service and IBM. The result is intended to be an open standard for packet data communications on cellular systems. CDPD will provide high-speed (19,200 bps) wireless packet data transmission and wireless circuit-switched data using standard modems, as well as capability to transfer to voice mode. It will make use of common circuits and components.

When cellular systems were first installed in the early 1980s, they were designed to work with mobile phones having a power output of 3 watts, traveling mostly on major highways. Over the years, systems have been expanded to work with handheld units with a transmitter power of six tenths of a watt. Coverage has been increased to include in-building coverage in most major metropolitan areas.

Since wireless data users and cellular phone users have the same requirements in terms of wide area coverage for low-powered handheld devices, it seems logical to use an existing infrastructure. Further, the cellular system is connected to the wired telephone network. Access to and from devices through wired systems is easily accomplished.

CDPD specifications, in addition to addressing the specifics of CDPD data transmission, allow for a choice between CDPD and circuit-switched data. In the circuit-switched data mode, the user merely adds modem software to the base cellular technology and makes a data connection that is the same as a wireline connection. The user can elect to switch to the packet mode and access CDPD, or to access the system for voice contact by adding an analog interface, a speaker, and a microphone.

System Principles

The CDPD consortium specified objectives that are designed to make CDPD the best choice for both voice and data. Among these are:

- CDPD will have no impact on voice capabilities, quality, or capacity.
  Data and voice systems are to be autonomous. Packet data systems will make use of "idle time" unusable to trunked voice systems. In any anomalous collision between voice and data, voice has priority.
- CDPD will leverage the existing infrastructure investment.
  No changes are required to existing voice system equipment. Existing antennas, cell sites, transceivers, switches, and switch software are not impacted by the addition of CDPD since CDPD is designed as a transparent overlay.
- Minimize market entry costs.
  Uses infrastructure in common with voice systems. The bulk of the system cost is for data transceivers, with a minimum of one per site being required, and that cost is about the same as adding one additional voice channel to a cell. All non-radio elements of the system can be built on general purpose computers.
- Enable a broad, flexible range of subscriber equipment.
  Both "things" and people can make use of CDPD. Things such as vending machines, utility meters, point of sale devices, and vehicle location systems can make use of CDPD. Anyone carrying a Personal Communicator or other mobile computing device can use CDPD.
- Make every unit capable of voice.
  The electronics are already in place for voice. All that is needed is to add a handset. This makes every portable computing device a cellular phone.

These points are the stated objectives of the CDPD group—all of which sound like viable objectives that will help make CDPD the accepted method to send voice and data.

CDPD Principles

The basic principle behind CDPD is that in any given cell there are multiple radio channels—all of which are not normally in use at the same time. For example, if a cell site has a total of twelve channels, it could handle twelve simultaneous voice calls. The chances of twelve voice callers being in range at any one time are assumed to be slim. If a someone within that cell wanted to send data using CDPD, his or system would request a channel and "hop" onto it. Since voice has priority, data only occupies a specific channel for small, specified periods of time and then "hops" to another idle channel. The CDPD system can, according to the specification, hop faster than the voice system can put a channel into use.

Why Hopping?

The reason to have data "hop" from channel to channel is that CDPD air time will be billed at a rate less than voice time. Therefore, voice must have priority in the system if service providers are to incorporate the CDPD system. In reality, the system as it is currently being implemented does not provide for data capability on all voice channels. A careful reading of McCaw Cellular's comments regarding CDPD—and McCaw's time frame for roll-out—reveals that the plan calls for adding a single data channel for most cells. Additional data channels will be added to cells with high demand.

The intention of CDPD is to offer viable communications for both the user and the cellular provider. Most of the provider's revenue comes from voice communications, so there must be no degradation to the voice system. Trade-offs necessary to accomplish this make combining voice and data over existing cellular systems complex.

Installation

The first CDPD system to be installed in the McCaw Cellular system was in Las Vegas. According to McCaw, Las Vegas was chosen because it is easy to get to and inexpensive housing is available for the engineers and technicians. Further, since Las Vegas is flat, fewer cell sites are needed than in many cities.

The Las Vegas installation, however, used dedicated data channels rather than data-over-voice channels. There was a very practical reason for this. The roll-out of the system coincided with the Comdex trade show. There are so many handheld cellular phones in Las Vegas during Comdex that CDPD transmissions—that have to defer to voice transmissions—in all likelihood would not have been able to get a channel!

Pervasive Data?

Again, a key difference between CDPD and other data transmission methods is that CDPD is a technology that must be embraced by a number of companies if it is to be universally accepted. RAM and ARDIS systems are one-company nationwide networks that have built-in common command and control functions. CDPD, on the other hand, will be implemented by a number of cellular carriers—but not *all*. CDPD may not be available in all areas or on the system selected by the user. If cellular voice-roaming performance is any indication, there will be inconsistencies in the way roaming charges are handled (how much is charged back to end users, and whether they can roam in an area without making pre-arrangements).

Unless CDPD systems become as integrated and as seamless as RAM and ARDIS are today, and as Mtel's NWN and others will be in the future, CDPD will not become a major contender in the wireless data marketplace.

Enter AT&T

Developments in late 1993 could help CDPD obtain the status of a wireless data contender. AT&T first invested in McCaw Cellular and then bought most of McCaw. AT&T's interest in McCaw goes beyond CDPD, but with AT&T's clout, financial backing, and its network expertise, CDPD's chances are better now than before the AT&T involvement.

The CDPD group has not addressed questions relating to the interconnection of the CDPD "pipe" to the information infrastructure. They seem, instead, to consider CDPD merely a pipe. Since the consortium is made up of cellular phone carriers, it is not surprising that their vision is to provide a pipe and hand off the interconnection logistics to others. It may be that AT&T's influence will change the emphasis on CDPD from a technology implementation to a wireless data system—complete with interconnection and advanced management services.

CDPD Conclusions

CDPD technology is based on the premise of modifying an analog system (the existing U.S. cellular system) to handle digital data. It remains to be seen if it will become nationwide in scope, and the technology is more complex than it needs to be to send and receive packetized data bursts over RF channels. If CDPD is successful, it will be because it has AT&T behind it.

ENTER MTEL WITH NWN

The next contender in the wireless data world is a company that already provides interconnected one-way messaging services in the United States, Canada, Mexico, Hong Kong, Singapore, and soon Indonesia, Malaysia, and all of Latin America. NWN (Nationwide Wireless Network) will provide internetting messaging service through its Global Messaging Network.

The paging service is run by SkyTel Corporation, and the parent company is Mobile Telecommunication Technologies Corporation. Most people refer to this company by its NASDAQ abbreviation—Mtel. Mtel is an interesting company that has big plans in the wireless data arena. It believes it can build on the experience gained in creating one of the most sophisticated one-way messaging and paging systems in the world. For an investment in the $100 million range, Mtel plans to become one of the premier two-way wireless network suppliers.

Setting the Stage

SkyTel provides both numeric and alphanumeric messaging services on two nationwide channels (931.9375 and 931.4375) and has successfully implemented 931.4375 for use in many other countries to permit seamless international roaming. The frequency is rapidly becoming a *de facto* standard for global one-way messaging.

In November of 1991, Mtel filed a Petition for Rulemaking with the FCC for the adoption of rules and policies that would allow Mtel, if licensed, to create a two-way nationwide wireless network in the U.S.

In April of 1992, the FCC awarded Mtel an experimental license (along with four other companies) to test its technology for a two-way wireless messaging network. These tests were to take place in a small sliver of spectrum in the 900-MHz band (930–932 MHz) which was set aside by the FCC for advanced paging technologies. In June of 1992, Mtel submitted detailed technical data demonstrating the viability of its Nationwide Wireless Network (NWN) technology based on field trials.

This was quickly followed (in July) by the granting of a tentative "Pioneer's Preference" status in recognition of Mtel's pioneering work in wireless messaging technology. The Pioneer's Preference category was established by the FCC to encourage technology developments. It also ensures the developers of new technology that they will be able to make use of it. (It should be noted that Mtel was the only company out of thirteen applicants that was awarded Pioneer's Preference because, as the FCC stated, "the company's proposal showed particular innovation and merit.")

The tentative Pioneer's Preference status of Mtel is being contested by several other companies that were passed over. This type of request for reconsideration, with possible further legal action if the FCC does not reconsider, is not unusual in such circumstances. It is doubtful that Mtel will lose its Pioneer's Preference standing.

Pioneer's Preference status is of great value in today's wireless world. First, the FCC and Congress will most likely require others that follow in the PCS business to go through a spectrum auction process that will raise money for the General Fund. Second, it gives Mtel a head-start on its competitors—except for RAM Mobile Data and ARDIS, already on the air, and CDPD, that makes use of existing cellular channels.

In December of 1992, Mtel and Motorola entered into an agreement whereby Motorola agreed to develop equipment for the NWN demonstration system to be built in the Dallas area. In January of 1993, Mtel submitted a progress report to the FCC summarizing the results of field trials held at the University of Mississippi. It also announced a $6 million investment from Kleiner Perkins Caufield & Byers (KPCB), a noted Silicon Valley venture capital firm.

In May, Mtel completed its testing of the Dallas demonstration system and submitted field trial reports to the FCC that further demonstrated the commercial viability of the technology. In June, the FCC confirmed its Pioneer's Preference for Personal Communications Systems award to Mtel—approving a two-way Nationwide Wireless Network (NWN) that will use a single 50-KHz channel in the 930-MHz band.

Others Take Exception

Why are companies willing to spend time, money, and effort to try to block the Mtel license? The answer has to do with timing. Remember that RAM and ARDIS are already operational as nationwide data systems and the CDPD consortium has committed to being operational sometime in 1994. Under a Pioneer's Preference license, Mtel becomes service supplier number four (or five, depending on the impact of Metricom, see below). Others will have to wait for the other 10 channels to become available (most likely as part of the government's new frequency auctioning plans), or for the FCC to pass rules for the 1800–2200-MHz band. They will have to wait further while the rules are contested and existing microwave users are moved to other areas.

Furthermore, those doing all this waiting will have to *fund* the moving of current microwave users.

Mtel's NWN will, however, be the first with a nationwide common frequency allocation for two-way messaging. And Mtel will have a good head-start over the Regional Bell Operating Companies (RBOCs), and over MCI, Sprint, and others eyeing such wireless communications systems.

Thus, there is a great deal of inherent value in the Mtel license. If NWN really is up and operating by 1995, the timing will be just about right, considering technological advancements being made in the communications and computer hardware side of the business.

The Goal

Mtel's goal is to provide two-way wireless communications with coverage in a minimum of the top 300 markets in 1995. This system will provide "location independent" service. (It will not be necessary to know where the sender or receiver is located—the message will be delivered.)

Key offerings are to include two-way messaging (including e-mail extension service), acknowledgement paging on-demand, broadcast information service, database access, transaction services, and fixed point services (such as Supervisory Control and Data Acquisition, SCADA).

In addition, the NWN Operating Center (NOC) will be a server farm available to other custom application developers for tighter integration of wireless capabilities with their applications. NWN is designed to integrate with other standard servers like General Magic's Telescript engine or other offerings from other vendors.

NWN is also the first wireless vendor to offer a virtual wireless network for integration with custom providers or public communications networks.

Again, the system is built around a single outbound nationwide 50-KHz channel and will use a data rate of 24,000 baud. Inbound transmissions (portable back to the system) will use the same channel and deploy frequency reuse in a given metropolitan area. The data rate for return messages will be 9,600 baud, with this lower data rate facilitating longer portable battery life.

The outbound channel, with its wider bandwidth, will be able to support 24,000 baud while the 12.5-KHz inbound channel will support 9,600-baud transmissions with frequency reuse.

Outbound vs. Inbound

Except for its 10x bit rate and frequency reuse from market area to market area facilitated by its return channel and centralized global subscriber location database, the NWN system will be similar to the SkyTel system. Its differences will permit location services, interactive information and transaction services, and a broad variety of tariffs, including metro-area only.

Since the heart of the system is a master computer that directs calls using the portable's address code, it is a simple matter to offer customized services. With NWN, Mtel can offer similar options and will, therefore, be able to offer a multiplicity of billing options. For a fleet of handheld units, some that need only regional communications and others that need coverage across the nation, the options can be programmed in a matter of minutes.

There are advantages to Mtel's strategy of using many more receivers in each area than transmitters. For inbound transmissions, it means that handheld devices can use lower-powered transmitters (saving on battery life, size, and weight). And since each handheld device will only be "seen" by a few receivers, the system will be able to "hear" more inbound messages per area.

Each area is covered by multiple transmitters operating in concert with one another, using a technique called "simulcast." For each transmitter, there are multiple receivers.

In actual operation, a handheld device would receive a message sent over the network and would respond to the message via the inbound channel. If a user were in motion during the transmission or reception of a message, the outbound message segments (packets) might be received by different devices—but the entire message would be "assembled" by the system before it was forwarded to its final destination.

System Capacities

Mtel's NWN staff is optimistic about high capacity, and projects it into the millions. They further feel that capacity can be added on an as-needed basis using techniques such as higher transmission rates and dynamic zoning techniques.

System Assessment

The technologies employed by NWN are well developed. The pilot program in Dallas successfully demonstrated the capabilities of NWN. As the roll-out continues, there may be some technical "tweaking," but the basic system has proven to be a combination of technologies that should work as advertised.

How the system will handle portable field units that can "hear" the base transmitters but are out of receiver range is unclear. The long-term answer, of course, is that NWN will add receivers to such areas as they are identified. The short-term problem may be, however, that some handhelds may not always be within receiver range. (This is not an NWN-specific problem. Radio coverage and the matching of transmitter and receiver coverage is not an exact science since radio waves bounce, are reflected, and otherwise misbehave.)

A second concern is with capacity and how NWN will provide cross-connections to other services. At this point, NWN is dealing with theoretical channel-loading data (as are *all* wireless data systems), and there is no way to know exactly how many customers such a system can serve on a continuing basis. This concern is not about overloading as much as it is about system delays in delivering time-critical data. Anyone experienced with wired networks knows that at some point loading reaches a critical level. For every new user, system access time is degraded by some percentage.

Early users will probably find NWN to be fast and responsive to their needs. There will only be a problem if as the user population grows, a difference in system performance is noticed. With the SkyTel nationwide paging system, the time it takes to deliver a message varies. Most of the time, an alphanumeric message is sent and received by the pager within one to two minutes. NWN anticipates the same quick message delivery.

Final Comments Regarding NWN

The technologies, caliber of the people, and the existing service offered by Mtel are impressive. The SkyPage system is robust, works as advertised, and is reliable. NWN should perform as advertised and the experience gained by Mtel with SkyTel will go a long way toward making NWN a reality within the specified time frame.

NWN has a decided advantage over others entering the mobile data market during the next few years. It already knows that the outbound system works—it has established the transmitter sites needed to provide wide area coverage. (Site identification, selection, and acquisition can be *very* costly in both time and money.)

NWN, as planned, is a data-only network. This is a plus. It is being built as a digital transmission system from the beginning. Another plus. And it appears that Mtel will have the financial strength needed to carry out full implementation of the system.

NWN has the potential to become a major wireless data network in the U.S. How well the promise is executed and how well the system is structured to work in conjunction with existing public e-mail systems, information suppliers, and in-house wired network systems will in some part determine what percentage of the total available market NWN will command—and it will be two years or more until it is fully operational.

What Mtel has is what appears to be a well thought-out and executable technology. It has a clear vision of what it will be able to offer users and it is working with a number of major partners in many areas to make sure all the technical and informational structures its system will rely on will be in place. In short, it appears that NWN will be successful and that it will be one of the survivors of the war between mobile data providers that is heating up.

METRICOM: AN UNLICENSED CONTENDER?

Up to this point, the systems presented have been either already licensed by the FCC and operational—existing two-way radio systems, SMRs, RAM, and ARDIS—or have been licensed and sanctioned by the FCC—CDPD and NWN.

Beyond these are many potential contenders that are watching and waiting for the FCC to release additional licensed channels or to hold license auctions. One new contender is proceeding with plans for metropolitan (and eventually nationwide) wireless data networks. It does not have to wait for the FCC to act because its system uses a portion of radio spectrum set aside for low-powered, unlicensed systems.

The company is Metricom, Inc. The technology is "spread spectrum," presently using frequencies in the 902–928-MHz band. This band is shared with other types of systems and it is legally used for unlicensed voice and data networks on a non-interfering basis. (As part of the Emerging Technologies proposals, the FCC has allocated an additional 20 MHz of spectrum in the 2-GHz band for unlicensed systems. Metricom technologies can also use this band.)

The central core of the Metricom technology includes a patent-protected, geographic addressing scheme that enables radios to communicate directly with one another. This peer-to-peer communication capability means that data from any point in the network can travel to any other point in the network without going through a "central intelligence" location.

Such technology permits redundant communication paths, immediate access to the network from any location, multiple simultaneous application capabilities, and, most important, the ability to truly automate operations through the implementation of "continuous intelligence" architectures. The technology developed by Metricom is about to become available to the world of non-dedicated wireless networks.

What This Means

According to Metricom, its design criteria for wireless data networks are based on what it believes customers really want—high capacity, low cost, and ease of use. Metricom's latest products are designed to meet these criteria. Its system operates at an RF data rate of 77,000 bps, which translates into a user rate of 50,000–60,000 bps.

The system consists of a series of radio transmitters and receivers installed as micro cells to cover a given area. Each micro cell has a capacity of 163 channels, and can handle up to 200 users per site (according to Metricom). Before delving into the specifics of how the system works, perhaps we should look at the most impressive aspect—the costing model. At present, Metricom plans to charge users a flat monthly rate—not dependent on the number of messages sent and received. Rather, the charge depends on the speed of the data over the network.

An example of this billing structure was provided by Metricom (not as final costs, but as target numbers it believes it can offer users while still making a profit). These examples are:

Data Throughput	Monthly Charge
2,400 bps	$2.95
14,400 bps	$9.95
56,000 bps	$19.95

It is these low prices, along with the low cost of system implementation, that has the rest of the wireless data network vendors taking a look at Metricom and what it says it can deliver. If this price structure holds true, Metricom will certainly have a positive impact on the cost of wireless messaging—at least in the metropolitan areas it plans to serve.

Metricom can offer low prices because its systems costs are extremely low. Since the individual transmitter/receiver units are self-contained and low-power, the cost per micro cell is in the neighborhood of $1,000 each (compared to cell prices of from $5,000 to $30,000 or more per site for other systems).

This makes Metricom's projected systems costs the lowest in the industry. Based on the number of radios needed in a given area, the cost per square mile of coverage would average around $50. It would only cost about $10,000 to provide coverage over a 300-square-mile area. This translates to between $5 and $20 per user in densely populated areas.

The Heart of the System

It is important to make a distinction between the Metricom offering and those of other wireless vendors. First, the Metricom system is strictly a wireless data pipeline. ARDIS and RAM systems also function as pipelines, but with their central intelligence hubs, they also perform system routing and other functions that require massive amounts of computing power. The intelligence built into the Metricom system is "network" intelligence. That is, it provides communications between nodes so that a message is routed from one point in a network to another. Until a unit signs on and directs a message to another unit, the network doesn't know about it. Once the unit connects to the network, the system can route traffic to and from it on a point-to-point basis.

Another difference between the Metricom system and others is that it uses what it refers to as a "mesh" network, and makes use of asynchronous frequency hopping, using spread spectrum technology. The advantage of this type of system is that each of the radio nodes is "smart" and each "knows" what other nodes are around it and which ones it can communicate with. (A "node" consists of a radio and intelligence.)

Mesh Technology

For a better understanding of the Metricom mesh approach, refer to the diagram below (Figure 7-2) showing the three types of possible networks. In a trunked channel system, there is a command and control system that "listens" to all the units and switches users to one of the channels when a request for use is processed. (Cellular phone systems can be considered trunked systems, since a command and control channel tracks all the users within a given cell and then switches the cellular phone to a specific channel before providing service to the unit.)

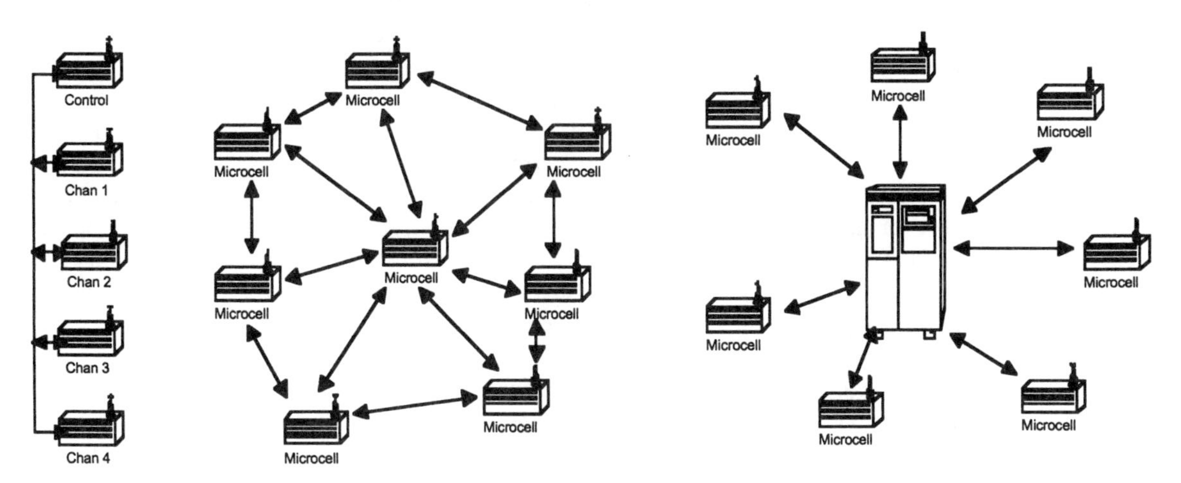

Figure 7-2. Metricom's "Mesh" Network

"Star networks," such as RAM and ARDIS, require all traffic to be passed to one or more central switching points before being connected. A star network is like the Federal Express network of wireless. When Federal Express picks up a package, it is sent to Memphis (or another regional center) where it is sorted and placed in an outgoing bin. Then it is loaded and transported to the Federal Express office nearest the recipient. There it is loaded on the truck covering the route that includes the recipient's address and is delivered.

Using this model, the Metricom system would make it possible for the truck that picked up the package to hand it off to another truck that might hand it off to yet another truck that would then deliver it to the proper person—without having to go through a central distribution point. A further difference is that the Metricom system provides point-to-point communications between handheld units without requiring them to access the network.

In the RAM and ARDIS models, to get a package next door, the package would have to have to go to the Federal Express regional sorting office to be placed back on the same truck it was picked up by, and delivered the next day. The Metricom system would be like walking across the street to deliver the package—getting it to its destination faster while not costing anything in the process. (The network is not used in direct unit-to-unit communication.)

In Practice

With the Metricom system, if two students on the same campus or two people in the same office wanted to exchange information and their devices could "see" each other over the wireless data path, they would not be switched onto a network that would send the message to a central control point and back out. The units would merely communicate directly with each other. When one of the units moves out of range of the other, the network would take over, relaying the message from the first unit to a node, to the second unit. If the second unit moved across town, the network would send the message from node to node until it reached the node closest to the second unit. The closest node would deliver the message.

When two units are communicating directly with each other and not making use of the network, there is no charge for the service. Two people within range of each other, or a desktop computer and a handheld, could exchange messages, files, and information all day long and not incur a single network charge. This is not true for any of the other systems. In fact, if you had two units outside the network coverage area of RAM, ARDIS, or NWN, both would be useless. On the Metricom system, they could at least talk to each other.

For many of the handhelds coming to market, the vendors have to design in a RAM or ARDIS radio *and* some other form of direct connection (such as a serial link, an infrared link, or another wireless transmitter and receiver—all of which add direct cost to the device). Devices built to work on the Metricom service need only a single transmitter and receiver.

Node Addressing

What makes all this work is that each node making up the network has a geographic designator or name. It "knows" where it has been installed and it "knows" the geographic addresses of all other nodes it can communicate with. When a node receives a message that is in transit (not intended for a unit at that node), it looks to see which node the message should be passed to next. Since each node hears multiple nodes, and since nodes can talk to each other via various paths through various nodes, the path chosen will be the best path available at the moment.

The Importance of the Metricom System

The Metricom technology is robust. Using "smart nodes" that are inexpensive ($1,000 each), the Metricom system makes a viable option for a number of applications. It can be used to provide wireless LANs inside a building, on a campus, or to cover entire geographic regions. One idea offered by Metricom is to build out its systems in a number of regions and interconnect them through some other wireless network, land lines, or even the Internet as the backbone.

The significance to a user is that with built-in peer-to-peer computing, a single radio installed in a portable computer can provide a wireless connection between the portable and a desktop system when the two units are in proximity to each other, can provide a connection to a company's LAN through a node attached to the LAN, and provide out-of-building coverage using the same unit if the regional area is so equipped.

Two units equipped with the Metricom system can communicate with each other without using the network and incurring charges. The Apple Newton MessagePad was designed with infrared capabilities, but unless one Newton is within a few feet of another, the system will not work. With Metricom-equipped units, information can be exchanged between *any* two systems.

Shortcomings

As with any potential system, there are some unanswered questions. In the case of Metricom, there are two areas in which there are either questions or concerns. None of these have to do with the core technology that is well thought out, protected by patents, and stable. The areas of concern have to do with frequencies and regulations.

As far as the frequency issue is concerned, the present Metricom system makes use of the 902–928-MHz band that is shared with many different types of users. Some of these users are licensed and, therefore, protected when it comes to creating or receiving interference from unlicensed users. In this particular band, unlicensed users rank number five in the pecking order—meaning that four groups of users are protected above them when it comes to interference issues. Also of concern is the emergence of this band as the new portable, in-home wireless phone band (also unlicensed), as well as garage door openers, remote control devices, and even low-powered walkie-talkies. The FCC is also considering using this group of frequencies for vehicle location, although it may be several years before any action is taken in this matter.

Metricom, realizing the limitations and potential for over-crowding in this band, has begun making plans to move up in frequency. As mentioned previously, the FCC has taken some actions to make the 1.8–2-GHz band available for unlicensed wireless networks and there are two additional unlicensed bands available. One of these is from 2400 to 2483 MHz (2.4 GHz), while the other is much higher—starting at 5.7 GHz. Based on all this information, we do not see the frequency issue as a hindrance, only one that needs to be properly addressed.

The next issue has to do with regulations. At present, there is nothing in the FCC rules or regulations prohibiting Metricom from providing its system on a regional or even nationwide basis. However, it is conceivable that some other company may raise the issue of using unlicensed channels to compete with licensed operators who have had to deal with the FCC and are required to answer to various states' public utility commissions. Even if Metricom is forced to limit its activities to campus-style installations, there are more than enough of these to provide a good revenue stream. On the other side of the coin is that the technology could also be employed using licensed channels. So, again, this issue is not of great concern.

Final Comments on Metricom

The opportunities for Metricom are enormous. It offers a different and robust network architecture, and its system provides clean and seamless access to users at a low cost. Metricom is a force for other wireless vendors to contend with. Metricom will have to be astute and clever when it comes to making alliances with others. How it structures these alliances will determine how big a player it will be in the wireless network arena. Metricom has some exciting technology, a staff that appears to be well versed in implementation strategies, and a growing following of believers.

Metricom can surely become one of the most significant players in the wireless data game, but it will have to work hard and smart to avoid being squashed by the big boys.

8

Recapping the Services

In summary, the basic types of wireless data services are:

- Paging and One-way Wireless Messaging
- Data Over Existing Two-way Voice Radio Systems
- Specialized Mobile Radio (SMR)
- Two-way Data-only
  - ARDIS
  - RAM Mobile Data
  - NWN (future)
  - Metricom (future)
- Data Over Analog Cellular
- Cellular Digital Packet Data (CDPD) (future)
- Personal Communications Services (PCS) (future)

PAGING AND ONE-WAY WIRELESS MESSAGING

EMBARC

The Motorola nationwide messaging service provided by EMBARC is different from the nationwide paging services that offer messaging. First, the EMBARC service has no "pagers." Motorola chose to differentiate its service and not to compete with its many paging service customers. The only receiving devices that can be bought or leased for the EMBARC system are designed to be used in conjunction with a computer and will not function as standalone pagers.

One of the most popular devices for the EMBARC service is a receiver and cradle that has been designed for HP 95LX or 100LX palmtop computers (Figure 8-1) to slide directly into the cradle and automatically make a connection. Any message that is then sent to the receiver can be displayed on the computer. EMBARC receivers *will* receive messages when not connected to a computer, but since there is no display, they *must* be connected to a computer of some type to be effectively received.

Figure 8-1. *The HP 100LX Palmtop Computer*

EMBARC realized early on that along with messaging, it would also need to offer other services. Shortly after its system was implemented, it began delivering news headlines from *USA Today.* Users can thus select the types of news and information they desire. The EMBARC service also provides stock market and other information on a subscription basis.

One of the best applications for EMBARC is "one-to-many" messaging. Companies that have a number of sales, service, or other representatives in the field with whom they need to communicate—to convey updated pricing, delivery, or other time-critical information—will find EMBARC a valuable service. Using the EMBARC system, a single message can be sent over the system to an entire group at one time, saving both time and money. This one-to-many capability is one of the best uses of a service such as EMBARC.

SkyPage

Meanwhile, Mtel—the parent of SkyPage—is expanding its service. SkyPage is now available virtually anywhere in the United States, Canada, Mexico, and also serves many other countries. SkyPage has added many new features and functions including SkyWord, an alphanumeric paging service capable of receiving text messages; individual direct 800 numbers for SkyTalk, a voice-mail system that takes a message and then notifies a pager that a message has been received; and, of course, direct access paging.

SkyWord service also includes news broadcasts using the Reuters news service. Scheduled broadcasts are sent twice a day, and important and late-breaking news is sent within minutes of its being received.

Most recently, SkyTel (SkyPage) added a single-access 800 number for all its services. This makes accessing a pager or messaging receiver easier and more convenient. For example, a SkyWord pager is capable of receiving text messages, numeric phone numbers, and notification that a voice-mail message is waiting. People who want to send a message can access the 800 number and simply key in a numeric phone number, or they can be switched to a voice system where an attendant will convert their voice message to text that is then sent to the pager. They can also leave a voice-mail message, or, with a computer, they can access the system and send text-based messages. This service makes a pager available to more people by providing multiple ways of accessing it, and enchances the system's capability.

MobileComm

MobileComm, the paging and messaging service of BellSouth, is also a nationwide service. It competes directly with the SkyPage and EMBARC systems. Built around what it calls the M:I:M:E (MobileComm Intelligent Messaging Engine), the system is designed to permit users to access it by way of a telephone keypad, voice, or computer.

MobileComm offers full nationwide coverage including all fifty states, Canada, and the Caribbean, or will provide coverage on a region-by-region basis. It also offers four priorities of message delivery—*ASAP* delivers a message As Soon As Possible; *Standard* messages are delivered within an hour after they enter the MobileComm system; *Overnight* provides for a message received before midnight to be delivered by 6:00 A.M. the next day; and *Notify* includes an alert that a message is waiting, displays a portion of the message, and stores the body of the message for later retrieval.

MobileComm also offers "sender pays" messaging to help users keep the cost of their paging service to a minimum. MobileComm is connected to a large number of public e-mail systems and can provide notification that e-mail messages are waiting, and display header information identifying the sender of the message.

One-way Messaging

There are so many one-way messaging options—and these options are changing and adding services quickly—that it would be impossible to provide a chart comparing their features. The best way to determine which providers offer the best types of service for the requirements of a user is for the user to discuss his or her needs with each provider. Choices should be based on the type of service, ease of access, and the *total* monthly cost of the system. Some services charge a minimum monthly fee and then add charges for each message received, or for the size of the messages, or use some other formula. If such a service is chosen, it may be more difficult to estimate—and therefore control—the *total* monthly costs associated with the system.

Regional and Local Providers

In addition to nationwide systems, each city and metropolitan area has scores of paging and messaging service providers. Most Regional Bell Operating Companies (RBOCs) offer direct-dial paging services, and there are local and regional service providers in every area. The choices are many and the services and costs will vary depending on the coverage (range) provided and the types of services offered.

Systems exist where those wishing to page a user simply dial a specific number and then key in their own number; or systems that work in conjunction with a voice answering service. Some are accessible with a computer, and some offer many of the same services as nationwide providers. Selecting a one-way messaging service requires some investigation and a clear understanding of the user's requirements.

For example, users who *never* travel outside their prime coverage area can probably get the most economical service from a local or regional provider. If they occasionally travel outside their prime metropolitan area, using a nationwide service—but subscribing to a regional plan—may be the choice. With this option, when a user is outside the normal coverage area, he or she can ask the service provider to extend the coverage for a specific period of time. Users will be billed for this expanded service on an "as used" basis. Those who regularly travel nationwide should subscribe to a nationwide system. Thus, they will not have to be concerned about the areas of the country where the messaging receiver will work.

DATA OVER EXISTING TWO-WAY VOICE RADIO SYSTEMS

If a company already has a two-way radio system in place, and if sending data to and from a vehicle meets the needs of the company, it is practical to consider adding data modems and displays.

The cost of adding data terminals may, in fact, rival the cost of the radios already installed in the vehicles, but since the system is owned and operated by the company, there will be no ongoing monthly charges for the system's use.

The disadvantages of adding data on top of an existing voice two-way radio system are twofold. First, there is potential for interference to the voice traffic already on the system. Second, most such systems have been designed to provide adequate radio coverage to mobile vehicle units and, therefore, they may not be adequate for data reception and transmission from within buildings, or to and from those on foot.

Also, in most cases, the use of an individual captive data communications system will require special software to permit the mobile system to interface with the existing computers. Unless there is a programmer on staff who can handle the ongoing software requirements, the expense of hiring an outside software contractor could be the most expensive element in the system.

On the plus side, if the data system is used for dispatch functions, drivers will be able to receive more detailed information—such as specific directions, and even maps—but the volume of voice traffic will often be reduced. This will permit more mobile units to operate on the same system. And, of course, there will be no monthly fees.

Ideal uses for digital data systems installed on top of existing two-way radio systems include dispatch operations, data collection from vehicles, and vehicle condition monitoring. If drivers are equipped with handheld data-collection devices and, upon returning to their vehicles, they plug the device into the data port, the data collected can be forwarded to the office after every call, speeding up billing and inventory control, or increasing efficiency. These systems can be expanded to include vehicle location through the use of GPS receivers in each vehicle, and the proper software on the main computer.

The final advantage to this type of system is better monitoring of the fleet, coordinating of tasks, and receiving and processing of information collected in the field, without having to key in handwritten field reports.

SPECIALIZED MOBILE RADIO (SMR)

Since SMR services started out as shared two-way voice radio systems, most of what was said about data over existing two-way voice radio systems applies to SMR services. Several companies decided to expand the basic dispatch function of SMR systems by adding paging, direct-dial telephone access, and digital data services. These systems, exemplified by NexTel and Qualcomm, compete directly with paging, cellular phone systems, and wireless data networks.

Companies offering these services contend that they can, and will over time, offer all the services now available through cellular, paging, and wireless data communications companies, and, in the future, by Personal Communications Service (PCS) suppliers.

Test systems in Los Angeles and other major metropolitan areas are in process. Reservations about this type of service have to do with system coverage when communication is attempted outside of a vehicle or within buildings.

Because SMR systems were designed to provide two-way voice communications from and to mobile radios operating at relatively high power (typically 35–40 watts, compared to 3 watts or less transmitted by cellular telephones), these systems tend to make use of a few strategically located radio sites, usually on top of high hills or tall buildings. They were not designed to provide radio coverage to low power handheld radios, and certainly not to handheld units inside buildings.

However, the SMR industry believes it can compete directly with cellular, paging, and wireless data network providers and can populate their systems with users who might otherwise opt for one of the other services. These systems will provide a multiplicity of communications modes to users, but the majority of SMR users will probably be existing SMR clients who want to add functionality to their two-way dispatch systems. Those who are persuaded to use SMR systems in place of cellular, paging, or wireless data systems will likely be disappointed with both the level of service and the coverage provided by SMR operators.

In some cases it will make sense to use an SMR service provider—where, for example, a two-way radio system is already maintained. Most SMR providers charge a flat monthly fee that is dependent solely upon the number of mobile and base stations on the system, regardless of the amount of use.

This type of system would certainly be worth investigating for already-existing SMR customers who are happy with the SMR's coverage and service if the SMR is willing to work with them to assure that specific software requirements can be met.

TWO-WAY DATA-ONLY

ARDIS

ARDIS dramatically improved its product offering in 1993. First, its "seamless" roaming architecture was completed. An ARDIS subscriber anywhere within its nationwide coverage area can access the network. Next, it signed a deal with RadioMail so that a subscriber can use RadioMail over the ARDIS network. ARDIS also began marketing programs aimed at attracting individuals and small user organizations in addition to its historical customer base—large corporate users.

While ARDIS is a nationwide data network with a backbone and infrastructure that can "find" users no matter where they are in the system, it is still simply a pipe. Users who bought a Motorola InfoTac and an HP 100LX and subscribed to the ARDIS network service, would not be able to communicate with anyone, nor would anyone be able to communicate with them. To complete the communications link and enable the system to be useful, users would have to add at least one other connection that knew of their existence on the network and could, therefore, talk to them. Because most of the existing users are private company systems designed to provide communications among employees of that company, there is no way to "learn" their addresses and send them messages. Avis car-billing people cannot talk to a Hertz, State Farm, UPS, or IBM user of the network. In the case of company-specific systems, the ARDIS pipe serves as a private wireless network. Unless ARDIS enables communications across companies, the users don't even know who else is on the network.

Thus, to be able to communicate over ARDIS (the same applies for RAM), it is necessary to do one of three things:

- Install a direct telephone line (called an X.25 connection by the industry) between the user's own office computing environment and a local ARDIS connection point;
- Install another wireless modem at the office and connect it to the computing system; or

- Subscribe to RadioMail and use ARDIS as a wireless transport between the unit and the RadioMail system that, in turn, would provide access to most public e-mail systems and the Internet.

The first two options would allow a connection to a private computing system. It would be necessary to obtain software for the wireless mobile unit that would permit remote access to the system. Such software is generally obtained from ARDIS or from a third party developer. It is possible to connect computing platforms ranging from a mainframe with 3270 emulation, to a network running Novell's NetWare, to a single PC using a remote access package.

The choice of connection (wired or wireless) will depend on the number of mobile users, and the level of traffic you and your company will generate on the network. Users engaged in a pilot program, should opt for a wireless modem connection and avoid the ongoing costs associated with an X.25 connection. As the number of users increases, a point will be reached at which it makes sense to opt for a direct-wired connection between the system and the wireless network. This decision may also be contingent on the physical location of the company and the location of the computing equipment.

If the company is in a prime radio coverage area and the computer room is not heavily shielded, it is likely that a wireless modem with a built-in antenna mounted somewhere within a serial cable's length of the server will work just fine. If, however, the company is not within a prime coverage area, or if the computer room is underground or heavily shielded, the choices are to put a remote antenna on the roof or some other area, or to make use of a direct wired connection.

Two-way radio companies have been making use of remote radios and antennas for many years. At present, however, no remote radio devices are available for the wireless data market. Since the idea is a simple one, when demand for wireless connections to corporate computing systems increases, remote equipment will become available for this industry. The two-way radio model is that the radio transmitter/receiver is mounted in a penthouse or other protected roof location, and an outside antenna is erected. The radio is then connected to a remote control unit in the main operation area by standard telephone cable. In this manner, the radio and antenna performance can be optimized, yet the control can be maintained where it needs to be—even if such a command location is not conducive to radio coverage.

Direct Connection Limitations

Connecting directly to a computing environment makes sense and proves to be economical when the main focus of a wireless communications system is to exchange data internally within a company. Such systems are dedicated systems that do not have the capability to provide access to other types of services. If there are requirements for connections to outside service providers, there are two basic methods to accomplish such connections. The first is to make the connection directly from the central computing system, providing access through the wireless connection. The other is to make use of a service such as RadioMail or some other similar capability that will be offered over the next few years. The ARDIS and RAM RadioMail connections are identical, and many of the multi-connect options are also common to both networks. These options will be discussed at the end of the RAM review.

RAM Mobile Data

RAM first announced availability of its commercial service at the E-Mail World Conference in late 1992. By that time it had already structured a deal with RadioMail, believing that it could attract both company and individual accounts to its network from the beginning.

In mid-1993, RAM completed the build-out of its first phase of nationwide coverage, including the installation of some 1,300 base stations providing coverage to about 85 percent of the total U.S. population. RAM's system, unlike the ARDIS network, required a build-out of between 10 and 30 channels in each area served. ARDIS has a common channel nationwide and adds channels in each area as needed. RAM, on the other hand, decided to acquire and build out a multi-channel system from the beginning.

During 1993, RAM, with the assistance of RadioMail and others, offered its service not only to corporate users, but to individuals and small businesses. Given its early relationship with RadioMail, PSI, and others, RAM acknowledged that this market was going to be built in two very different ways—by capturing large corporate accounts, and by making it accessible to individual users who would launch wireless campaigns within their own companies and business communities.

Like the ARDIS network, the RAM system requires a wireless modem and radio transmitter/receiver. Most users on the RAM system use a device called a "Mobidem" built by Ericsson. Like the Motorola InfoTac used on the ARDIS system, the Mobidem unit makes use of specialized, but different, network protocols and is not compatible with most existing software available today.

In June of 1993, RAM, Intel, and Ericsson made a joint announcement that they would work together to provide wireless equipment and access which would be easier for the end user to buy and which would enable the use of many existing software applications. The first product to be introduced as a result of this cooperation was the Intel Wireless AT Modem, designed and built by Ericsson, labeled and marketed by Intel.

The companies involved hoped that if it was easier and less expensive to obtain wireless modems, and if they were compatible with existing applications, more users would be tempted to buy them and experiment with wireless communications.

These announcements were followed by agreements with Lotus, Microsoft, and AT&T EasyLink to provide Intel AT Modem-compatible versions of their popular mail packages. During the third and fourth quarters of 1993, these packages began finding their way into stores. At the same time, Intel initiated a program to train the retailers that would handle the products. But by the end of 1993, it was not clear whether this method of selling wireless connectivity would have wide appeal or if it is indeed possible to sell such products effectively through retail channels.

A Good Way to Learn

Regardless of how successful this program is, it provides an excellent opportunity to gain first-hand experience with wireless connectivity at a very reasonable cost, with a minimum of effort. If a company uses cc:Mail or Microsoft Mail, all it needs to get started with wireless e-mail is two of the Intel Wireless AT Modems (one for use with a portable, and one to connect to the server). These modems cost less than $800 each, and the software for either cc:Mail or Microsoft Mail costs less than $200. If a company wants to access the world of public e-mail systems, the Intel wireless modems and the AT&T EasyLink connection make the most sense. Here again, the cost of the software is not an issue. Once connected to AT&T EasyLink, a company will have access to any mail system that also has a connection to the Internet (not to be confused with an "Internet connection" that provides direct access to many of the services and information sources on the Internet).

No matter which of the three types of service a company chooses, for an investment of less than $2,000, it will have all the necessary wireless hardware and software. The monthly charge for unlimited service for the first three months is $75 per month (followed by monthly charges between $25 and $169, depending on how much the system is used). The AT Modems provided by Intel may not be the best way to access wireless services (see below), but they do work, and they will permit most standard communications-aware software to function across these services with little or no modification.

RadioMail

If a company does not have an e-mail system based on either cc:Mail or Microsoft Mail, there are other options for experimenting with wireless communications. The most popular of these is to establish an account with RadioMail. RadioMail can be accessed over either the RAM or ARDIS network and provides each user with a RadioMail address that is also accessible through any mail system that connects to the Internet.

The advantages of using RadioMail are that it not only provides users with a common, public e-mail address, it also provides access to world and business news, and soon it will offer access to many other services that should be attractive to end users. Further, with RadioMail's fax service, it is possible to convert an e-mail message directly to a fax message and send it to anyone who has a fax machine—even if that recipient does not have an e-mail address.

A key element of mobile computing is the element dubbed the "to-do" handler. The assumption is that when users receive an e-mail message, it usually contains some information that requires action to be taken. In the normal flow of business, this requires them to add the item to their "to-do" list and then address it at the appropriate time. However, if users have access to other communications channels, such as fax capabilities, they can handle many of these issues quickly and efficiently. For example, if they are traveling and receive a wireless e-mail message requesting general information, they would normally have to save the message and print it out when they return to the office. With the ability to send a fax directly from the RadioMail wireless e-mail, users simply "forward" the message to the fax machine in their office and the item is acted upon—they can continue working knowing that the message has been properly addressed. This is referred to as "instant" to-do handling—it is a great way to save time and effort. For the immediate future, many people will not have wireless e-mail, but they all have access to a fax machine.

Two E-mail Addresses

An inconvenience associated with using RadioMail and an internal e-mail system is referred to as the "two mailbox" problem. If users already have an e-mail address on their company's internal e-mail system, and they then subscribe to RadioMail, they have two mailboxes to check. It is possible to "forward" from a company e-mail address to RadioMail so all the e-mail can be accessed through the RadioMail mailbox. All of the internal e-mail that has been accessed remains in the inbox, however, with no indication of what has been read.

RadioMail, Lotus, Microsoft, and many others are working on solving this problem by adding the ability to "filter" e-mail so that users can pre-define which messages are sent to RadioMail and which messages stay with their local e-mail system. Handling two mailboxes is not a major issue and should not be a reason to wait to implement wireless e-mail. The solution is close at hand, and this problem will be solved in a time frame acceptable to most users.

In the meantime, the combination of RAM and RadioMail, or ARDIS and RadioMail provides a viable wireless e-mail connection for many users who want to access their e-mail wherever they are. The ability to access e-mail any time and virtually any place (within the major metropolitan areas) is a valuable asset that provides a "leg up" over their competition.

Nationwide Wireless Network (NWN)

Mtel, the parent of both SkyTel and NWN, believes that it will be able to begin offering two-way messaging and data services over NWN in 1995. Its service offering is distinctly different from that of either RAM or ARDIS. NWN is working with Motorola, Wireless Access, and several other companies to create a new generation of handheld wireless messaging or two-way pager units. The first of these will provide pre-programmed response messages such as: "I received your message and will call you later," or any number of different, short, responses.

It appears as though this service will be best suited for those who need both one-to-one and one-to-many communications. Many of NWN's first users might come from the SkyTel base of users who have experience with nationwide paging and alphanumeric messaging and are looking for the "missing link," the ability to acknowledge receipt of a message and to be able to respond to it.

NWN's system is ambitious, but the design concept appears to be sound and it should work as advertised. It is, however, not clear whether NWN is intended to compete with ARDIS and RAM for e-mail and desktop access users, or whether its primary purpose is to provide a new form of paging to existing pager users.

Metricom

Metricom is in the process of installing its unlicensed networks in a number of campuses and metropolitan areas within the United States. Metricom's unique network typology will enable it to provide PCS-like services, long before PCS providers will be able to offer networks.

Because of the technology used in the Metricom system, it provides for communications directly between portable units located within range of each other without requiring network (peer-to-peer) access, access through a network node to existing LAN systems from anywhere within a building or campus complex, and, (if the Metricom vision is achieved), access from anywhere in the major metropolitan areas.

The most important part of Metricom's network strategy is that it is the same as the PCS model—users will not pay for air or access time when using the system within their own facilities. Upon leaving their facility, they will be automatically connected to the network and will be using the portion of the network for which they will be willing to pay a monthly charge.

At this juncture in wireless communications, it is difficult to expand on the true capabilities of the Metricom system. It is possible that it will be deployed as a standalone system and, therefore, compete with ARDIS, RAM, NWN, CDPD, and PCS. It is also possible that Metricom might form alliances that would provide its technology and expertise to one or more of the other system providers. If so, it would provide local access to a nationwide system, making use of one or more nationwide system as an intelligent backbone to tie its networks together.

Due to increasing congestion on the band, there are concerns regarding the continued use of the 902–928-MHz unlicensed band. However, Metricom is aware of the potential for problems and is working toward providing its technology in the new 2.1-GHz unlicensed PCS spectrum and on the other two ISM bands (2.4 GHz and 5.4 GHz).

It is too soon in the evolution of the Metricom system to be able to provide specific comments regarding its implementation and the advantages and disadvantages of the technology. It is not too early to understand that of all the potential suppliers of wireless networks, Metricom is the one company that is able to support user devices which can communicate directly with each other and can also communicate over the network. In other systems, devices are captive to the network they use. If users are not within range of the network, they will not be able to send or receive data.

Data Over Analog Cellular

On the surface, using the existing analog cellular network would appear to be the best way to accomplish wireless data communications. Computer users familiar with attaching a modem to the serial port of their computer and then connecting the modem to a phone line to access remote information or e-mail might think that using an analog cellular network would be as simple as using these same components, including the computer communications software and the modem, and substituting a cellular phone for a wired one.

Unfortunately, this is not the case. Even though the cellular industry claims that up to 10 percent of its voice phone users are also sending and receiving data over their existing phone equipment, there are many who have tried and failed in this endeavor.

Even if a user were able to make a modem connection over a cellular service successfully, the costs to send and receive data would be much higher than those charged for a wireline connection. The difference is, of course, that cellular users must pay for "airtime." Every minute a cellular user is connected to a host computer through a cellular phone costs between $.20 and $.55 (depending on your service provider). Thus, a ten-minute call to an 800 toll-free number to send and receive your e-mail messages will cost between $2.00 and $5.50.

In reality, connecting to a remote computer over the analog cellular network is not as easy. First, the software on both ends must provide for some form of error correction to assure that the data is received properly. Second, the system must "understand" that if the call is transferred from one cell site to another, there may be a switching delay of up to 300 ms—long enough to indicate to the other end that the connection has been broken. The most successful users of analog cellular are those who settle for 2,400-baud modems equipped with special data compression and error correction schemes. Even these users admit that they cannot make a data connection every time they try.

The good news is that as cellular companies continue to become more involved in digital data and CDPD, their systems become more responsive to analog data. In the future, when all the cellular systems have been converted to digital voice (CDMA or TDMA), the process for sending data will most certainly improve. For now, however, sending data via cellular systems can be done, but this should not be one's only form of mobile data access.

Likewise, sending a fax in its native form is not possible on *any* packet data network at this point. Future advances will make sending and receiving fax messages over packet systems possible, but at present, there is no way to packetize faxes in their native format. This is one of the reasons most cellular carriers are promoting not just CDPD (see below); instead they recommend a combination of voice, analog data, and CDPD as the most attractive combination of solutions for the mobile worker.

Cellular Digital Packet Data (CDPD) Future Options

The use of CDPD should be considered as a "future" option when selecting the proper technology for a wireless data communications system. However, some areas of the country are already equipped with CDPD, and more metropolitan areas are being equipped each month. CDPD is an important technology, but it is still a future choice because it will take time for the cell sites in all the major metropolitan areas to be converted to permit the use of CDPD. Even more time will be required for the various cellular carriers to offer a nationwide seamless data system with a single bill at the end of each month.

Cellular analog voice systems have been in operation for more than ten years now, and users are still are not able to make calls from anywhere. Successful communications made outside the user's "home" area are charged a "roaming charge" that can run as high as $3.00 per day and $.90 a minute. The cellular industry has indicated its desire to provide nationwide CDPD coverage, and great strides have been made in this direction. However, not all cellular carriers are willing to spend the time and money involved in implementing CDPD until they see a demand for the service.

The cost of RF modems for use with CDPD is still in the $1,500 and above range. One modem, from Cincinnati Microwave, sells for under $300, but it is designed for CDPD use *only*—not for voice, analog data, *and* CDPD. It is best suited for point-to-point data transfers. There will be a huge market in this area for validating credit cards, monitoring the proper operation of a vending machine (the machine "tells" the dispatcher when it is empty or needs service), but for general public access, a CDPD-only modem does not make much sense.

Other issues must also be resolved if CDPD is to become a contender for mobile data usage. First, access must be nationwide and seamless. Second, the system must be an intelligent network that provides for easy access to users' own computing systems, as well as to many different types of information suppliers. Last, the existing rules governing the use of two phones with a single phone number must be amended.

The problem, briefly stated, is that each cellular phone is assigned both a phone number and an electronic serial number (ESN). This dual identification concept serves two very specific purposes. One is that carriers are assured that a subscriber does not use two phones with the same phone number, thus circumventing the monthly access charge. Another is that because both numbers must be verified, if a phone is reported stolen, it can never be used on any cellular system again. (Cellular phone thefts are almost non-existent, except by those who have learned how to illegally change the ESN).

As a result, if a user has a small handheld phone today and wants to take advantage of CDPD, he or she must either pay a second monthly access charge and obtain a second phone number for the CDPD-equipped unit, or get rid of the existing phone and use only the CDPD phone. Most users will not want to carry a complete computer and CDPD-equipped phone with them all the time. They will want to be able to carry their voice phone most of the time, and the CDPD phone when data transmission is important.

Users who are planning a wireless communications system in the immediate future should check on the latest status of CDPD-equipped cells in their area and the time frame for service availability for the other areas in which they will need coverage. They should also make sure to fully understand how the CDPD system works—including its possible limitations—before committing to a CDPD course of action.

CDPD will become an important part of the wireless scene, but to date, it has been oversold and underdelivered. A potential user must approach CDPD implementation with care.

Future Personal Communications Services (PCS)

The final choice for data—and voice and data—will make use of the newest set of frequencies to be allocated by the FCC, those for PCS systems. There is still a long way to go before the frequencies are available. Licenses will be issued through an auction process with the highest bidders being awarded licenses for one or more areas of the country. Systems must be developed and standards agreed upon, and then the actual systems can be constructed. During these activities, the existing users of this spectrum must be relocated. The cost for this relocation must be borne by those intending to provide PCS services.

The FCC also allocated 40 MHz of unlicensed spectrum for use within buildings and campuses. Currently, 20 MHz of this unlicensed spectrum is for voice use, and 20 MHz is for data. It is not clear how many licensed PCS providers will build their systems with the intent of providing both voice and data services, but since they will all make use of digital transmission techniques, most systems will be capable of both, should the service providers elect to offer both on their systems.

That the FCC has moved as rapidly as it has in allocating these frequencies underscores the government's belief that PCS is important to the future of United States technology companies and to the future of our continued technological leadership. There will undoubtedly be many delays along the way toward PCS implementation. The bidding process itself is a first for both the government and the industry. What follows the first round of auctions remains to be seen.

Those planning to make use of wireless communications systems should watch the continued progress of PCS, but they should also realize that it will take a few years before the capabilities that will be offered by the successful bidders are actually known. Fully functional PCS systems are not expected to be available for use until toward the end of 1996, or more likely, well into 1998.

PUTTING IT ALL TOGETHER

Viable options for implementing wireless communications are available today. More will be available in 1995 and 1996, and even more beyond that. Those who wait for all the pieces to be in place will be rewarded with fully seamless, go anywhere, do anything, wireless computing. They will find, however, that their competitors, who moved ahead before all the pieces were in place, have left them behind and have a decided advantage.

Implementing wireless computing in the 1994–95 time frame will require a higher level of understanding than it will in 1996–97. It is possible—and practical—to begin implementing wireless systems today. Chapter 9 provides guidelines for determining whether a company should be moving ahead with implementation and, if so, how to proceed, or whether it should sit out this round to wait for what is just around the corner.

9

A Guide to Implementation

Hopefully, the preceding eight chapters have provided enough information so that readers can determine whether this "new" wireless mobility will be of value to them and their company, or if it is merely interesting, and not worth their time and effort at present.

This chapter provides questions that need to be answered to determine the proper types of wireless communications systems for readers who are interested in entering the world of wireless mobility. The most important thing to remember is that this is not a "one size fits all" business. Some people may, in fact, be more productive with a simple paging receiver, while others may need complete, instantaneous access to a central information center or some external information source.

For those who want to delve further into the various technologies behind the communications links, a technology section follows in Appendix B. This includes a fairly detailed technical explanation of how many of the systems discussed work, why they use the channels or frequencies they do, and what their technical limitations are, if any. These technical issues are presented in such a way that readers with little or no technical background should be able to gain understanding and insight into the workings of the world of wireless communications.

SELECTION GUIDELINES

For readers who think wireless data communications can provide a new and expeditious way of getting information to the person or people who need it, and thereby increase their productivity, the following questions may prove helpful:

- How critical is the flow of information—can it wait until a person calls in, or is it of more value if it is delivered faster?
- Is interactive data communications necessary for the company?
- Can the information provided be equally useful if it is sent or received later?
- Does the company need a system where the information is sent and the person on the receiving end is notified that information is waiting? (RAM and ARDIS systems require the user-recipient to check into the system to retrieve the information. Paging systems send the data and notify the receiver with a tone or other form of instant notification.)
- Once the information is sent, is a quick response necessary? If so, does the response go to the same person who sent the message, or does it go to someone else?
- If the response goes to the sender, how will it be received, and how will the person be notified?
- Are one-to-many notification and/or information delivery required?
- How much equipment will the mobile person be willing to carry? (A field service person might be required to carry a handheld computer and radio device, whereas an executive might only need a pager that can be worn on the belt.)
- Does everyone on the mobile staff need to have two-way communications capability, or can a two-way system be mixed and matched with a one-way paging system?
- How much is the company willing to pay for this service?
- Once the transmission and reception of the company's prime messaging needs is accomplished, what else could be done with the system? Of how much value is this to the company?
- Are the people who are going to be using this equipment technically competent, or does technology scare them?
- What level of assurance is necessary to verify that the message was delivered; how important is notification of the message's arrival?

The answers to these questions will, for the most part, determine which type of system is needed. It may be that a combination of message paging and two-way interactive data will be the best approach for a user, or it may be that urgency factors outweigh the niceties of providing a two-way link between the office and the mobile worker.

START SMALL, GO SLOWLY

In any event, it is always best to start out slowly.

- Define the system objectives, implement a pilot system, and then measure its success against the objectives.
- Experiment with more than one type of system if at all possible. It is better to spend a few extra dollars up front than to place a large systems order only to find out that the concept does not meet the reality.

- Explore ways to minimize start-up and evaluation costs. For example, the Intel/RAM Mobile Data AT Wireless modems that are available in many computer retail stores for around $700 each provide an excellent way to determine if wireless communications will really benefit a user or organization. By buying a pair of these modems—and the software to access Lotus cc:Mail, Microsoft Mail, or the AT&T EasyLink service—users can quickly and easily install the beginnings of a wireless communications system for less than $2,000.

Following this course of action will render first-hand experience in the area of wireless data communications without requiring the user to do more than receive the system, be assigned an address, and experiment.

PRODUCTIVITY

Will wireless communications increase productivity? The answer will depend on individual particular circumstances, how critical it is to get information from one person to another, how much can be saved by implementing such an experimental system, and how much the people who will have to use it understand about its value to them.

Users who want to move ahead should choose the type of wireless system they think they want to implement. Starting with a small number of units, they should work against a pre-defined set of principles and objectives. If the problem cannot be defined, it cannot be solved.

BETTER OVER TIME

Ease-of-use will come with time and with people's understanding of wireless communications, their perception of the need, and, most importantly, the benefits they can identify. It will not be easy at first. As with anything that is worthwhile, however, users must be willing to try new ideas and explore new methods for getting the information to where it needs to be, when it needs to be there.

These are information-intensive times. Winners will be those who continue to understand the value of the information and to capitalize on that value. Those who do not perceive the need for wireless data communications will probably not be in the early-adopters camp—they could lose the advantage point to their competitors.

IS WIRELESS IN YOUR FUTURE?

In the next few years, as the information age moves from the desktop to the palmtop, as information becomes even more vital to a company's success and to each individual's success, wireless communications holds the promise of providing solutions that have not previously been available. Developing these systems will still be a long, slow process, but now is the time for potential users to start looking at the way in which they do business to determine whether wireless data communications will assist them in becoming more productive and more competitive.

Users should also remember that wireless communication will not get any harder than it is today; it will get easier. Those who try it and discover that wireless is still not ready for prime time should follow the large number of new devices and services that will be offered for this new medium, and be prepared to forge ahead—slowly at first.

Potential users should make sure to predetermine their goals and measure their successes and failures against these criteria. Wireless data communication is here and usable now. Hesitation might provide users with a better overall design of product next year, but they will be faced with the same initial question: "Does being mobile, and specifically wireless-data-communications-enabled, really help to provide better service to their clients, customers, or support people located within the same geographic area?"

Mobile computing is fun, it is a sound business practice, and it can save time and money—two things which, after all, are *really* in short supply.

IMPLEMENTING WIRELESS DATA STRATEGIES

Those considering a wireless option need to decide when they should proceed. Should they try a small pilot program now, or should they wait for CDPD, NWN, or some future technology?

The need for wireless access should be carefully examined. If a company is considering implementing a wireless system to explore its potential, invest slowly. If the examination indicates that a company may be able to save money, increase service capabilities, provide customers with better care, or otherwise offer an advantage over the competition, implementation should proceed as follows:

- Plan to start small. Do not try to implement all potential mobile computing and data solutions at once. Identify one major use for the system; start with that and build on it.
- Avoid running wired circuits to a provider's network. Make use of wireless connections from the host to the service provider's equipment whenever possible. A dedicated circuit can be added later if warranted by the volume of traffic.
- Seek professional assistance to determine the appropriate options. RAM, ARDIS, and a number of value-added reseller companies in the networking field offer such services, as well as an increasing number of consultants.
- Mimimize any disruptions to the existing IS system. Design a system that integrates into the existing service and "looks" to the system like just another connection. For example, if the other connections are 3270 terminals or terminal emulators, the first generation wireless system should provide terminal emulation as well.

- Do not invest more in wireless than the company can afford to write off or recoup in savings over the course of two years.
- Make certain that all the system's costs are known and understood. When making a decision to use a service that charges on a message-by-message basis, estimate the total monthly charges and track them closely.
- Invest in training for the system's users. Many systems that are technically viable fail because of a lack of user training or the failure to involve end users in the implementation process.
- Review, modify, and change. Work with the supplier to maximize system flexibility.
- Do not assume that one type of wireless system will meet the needs of all of a company's users.

SUGGESTIONS FOR THE SHORT TERM

When experimenting with wireless connectivity, following the suggestions and pointers below will minimize the potential for problems.

If a wireless modem is used with an existing portable computer:

- Set up and connect the system using *only* the cables supplied by the wireless modem vendor. These cables will be properly shielded and will provide maximum isolation between the wireless modem and the computer.
- In addition to testing the system in the store, test it in several different locations such as in a car, office, and any other locations where it may be used.
- If possible, design a case or device to hold both computer and wireless modem so that their relationship to one another remains constant.
- When sending and receiving data, look at the computer screen to make sure that it is not wavy when the transmitter is on or that it does not suddenly shrink in size when the transmitter comes on.
- If a cellular phone will be used as well as the wireless computing device, make a call while sending and receiving messages to make certain that the units do not interfere with each other.
- Experiment with the unit to become aware of typical times for the system to send and receive messages. Determine how soon the unit starts sending and receiving messages after it is set up. Determine how long it takes to send or receive a typical piece of e-mail or a file. While times can vary because of network activity, they can also vary because of radio interference between the wireless modem and the computer. Knowing typical times facilitates identifying when there is interference.

If a packaged unit is purchased:

- Make sure that the package is a standard configuration for the pieces and parts. (The "Road Warrior" kit offered by Ericsson uses an HP 95 or 100 LX and a wireless modem fitted into a case (Figure 9-1). Extensive testing has been performed to assure that they work well together. ARDIS has a similar package that uses the same computer and the Motorola InfoTac wireless modem.)

- If the package has been assembled and integrated by the company selling the devices, make certain that they have a knowledgeable person on their staff who can offer assistance if there problems are encountered.
- Experiment with the package as above.

Figure 9-1. *A "Road Warrior" Kit*

CONCLUSIONS

Wireless computing is here now and it can become a valuable and indispensible part of a business. It is not yet anywhere/anytime computing, but wireless access is certainly a valuable addition to the mobile person's resources. The costs associated with becoming wireless-enabled are not unreasonable, and the ongoing costs of normal operation are less than $100 per month.

The uses for wireless communications are many, and the reader can expect to learn of many more options as users' needs and desires are identified (Figure 9-2).

If users want pen-based systems, determine how much of their data must be received at the computer center in the form of ASCII text and how much can be handled by pull-down, multiple-choice menuing systems. While pen input is not well suited for ASCII text entries, it offers convience when coupled with multiple choice selection.

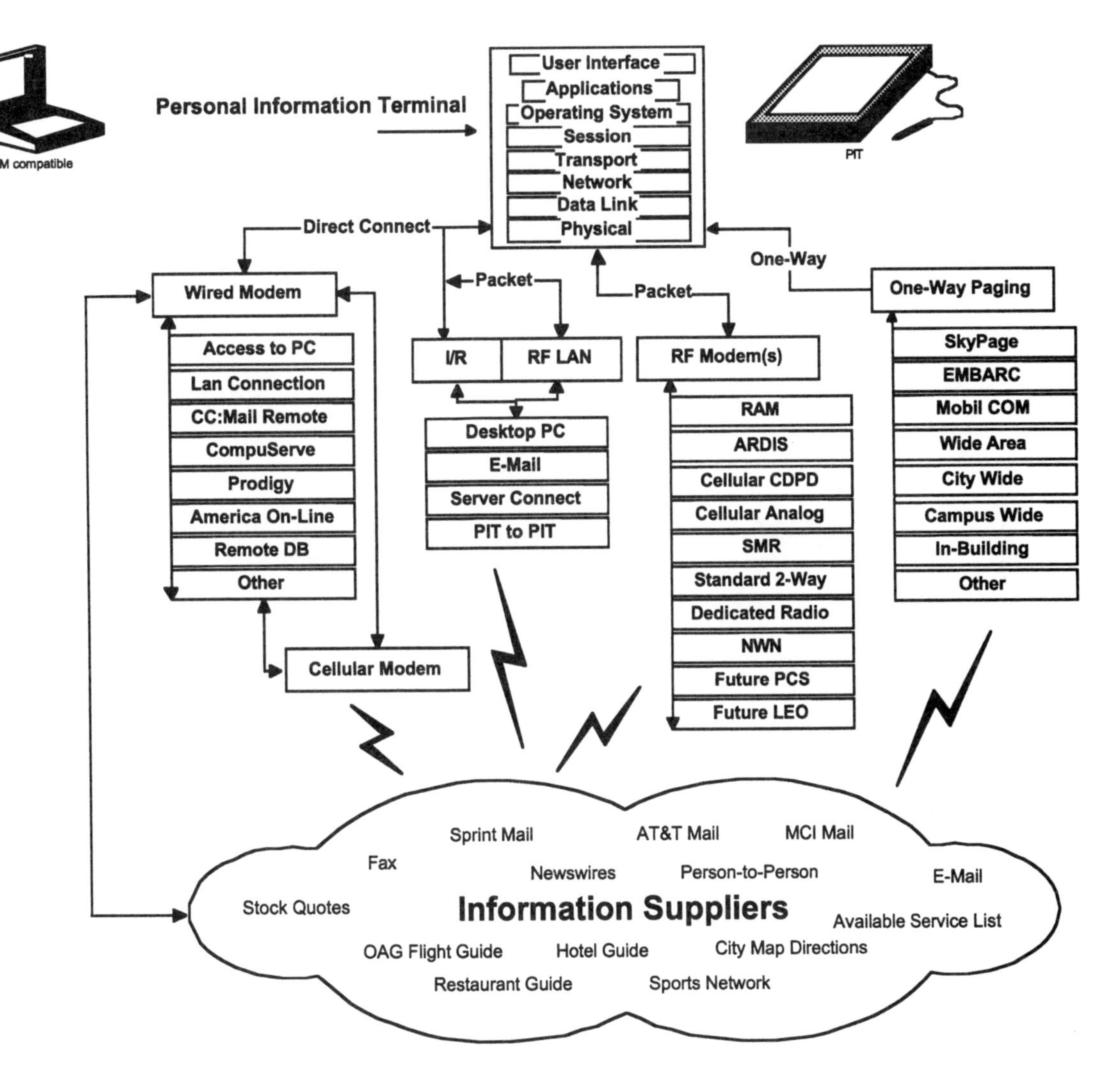

Figure 9-2. *Possible Connections to a Personal Information Terminal*

Above all, take time to understand the requirements of field personnel and the traffic they will generate as well as the level of accessibility they will need to complete their tasks efficiently.

Wireless data communication is convenient and economical today and it will become more so, and wireless data communication can provide access to those within a company who need to stay in touch. It is not a cure for controlling field personnel, and it is not a total solution. Providing wireless data access today requires a certain level of commitment for success; however, once implemented, users will wonder how they did without it.

Appendix A

Wireless Systems and Networks

SIMPLEX RADIO SYSTEM

When the frequencies were allocated to the business and public-safety radio services in the low band (30–50 MHz) and VHF band (150–172 MHz), they were issued as specific channels, or single frequency allocations. At that time this system worked because all communications services went from a base station radio transmitter/receiver unit directly to a fleet of mobile units.

Thus, a fire department might have been assigned a channel (frequency) of 154.415 MHz for its use. All of the vehicles and the dispatch center were then equipped with radios that would transmit and receive on the same frequency.

When a unit transmits, its own receiver is muted so that it will not hear itself or be subject to feedback, but any of the fleet within range hears the transmitting unit. After the transmission is complete, the "talker" releases the Push-To-Talk (PTT) switch. This turns off the transmitter and permits the receiver to hear a reply. This type of communication is referred to as "simplex" communications (Figure A-1), and it is probably the most familiar of all forms of two-way radio communications. It is the type used in the CB radio band and depicted in police shows on television.

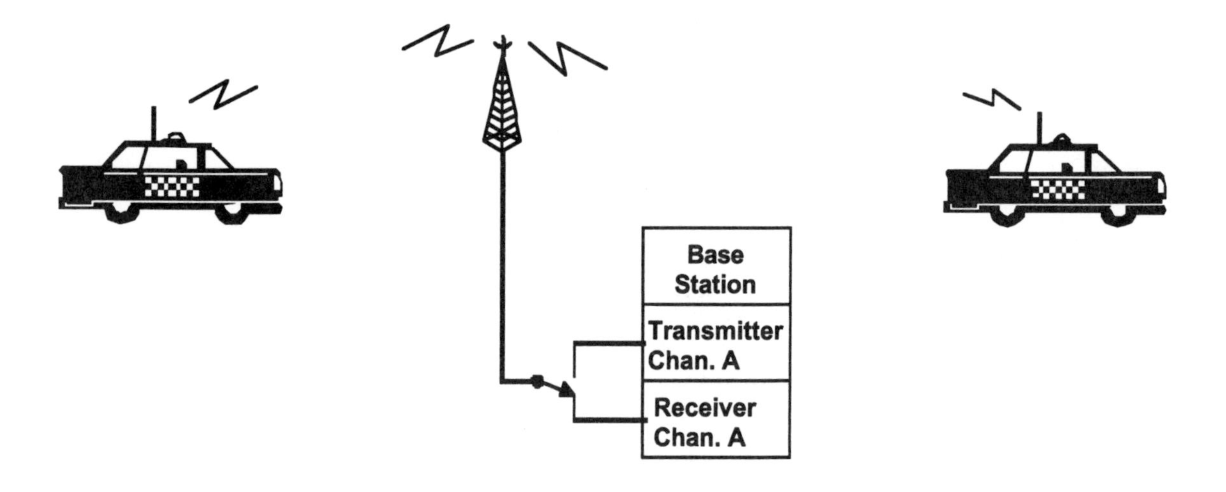

Figure A-1. *Simplex Radio System*

Obviously, only one unit can talk at a time. If more than one person is trying to talk, the result will be that neither person will be understood. This type of system requires some sort of coordination, which is usually imposed by the base station, or "dispatcher." In most cases, the calls are direct in nature—a dispatcher calling a specific unit by number, giving it information, and then listening for a reply.

The advantage of this type of system is that is it controlled. The dispatcher is using a higher-powered radio and a larger antenna than the individual units and can effectively coordinate use of the channel.

The major disadvantage of this type of system is that the range of the entire system is limited to the distance that can be covered from the base station to the mobile radio and back again. In flat countryside, this coverage area might be fairly good—up to 40 or 50 miles or more depending on the frequency used, the power of the transmitters, and the relative height of the base station antenna. In a hilly or populated area, coverage will be shortened by changes in elevation, buildings, etc.

Since the mobile unit's transmitter power and antenna are not as high as the base station's, car-to-car range is very limited in a simplex system. Generally, the range is only a few miles. If a vehicle in the northern part of the coverage area wants to talk to a car in the southern part of the area, it must rely on the base station operator to listen to one vehicle, repeat the message to the other vehicle, and then reverse the process.

Another disadvantage of this type of system is that since the two vehicles may not hear each other, there is a chance they will both talk at the same time, making it difficult or impossible for the dispatcher to hear either one.

REMOTE BASE STATIONS

The earliest method devised to increase a radio system's coverage was to locate the base station not at the main company office building, but at a location with a higher elevation. The base station and antenna systems were installed on the roof of a tall building, on a mountain top, or other high spot. A pair of telephone wires was then run from the base station to the dispatcher's office to control the transmitter/receiver by remote control (Figure A-2).

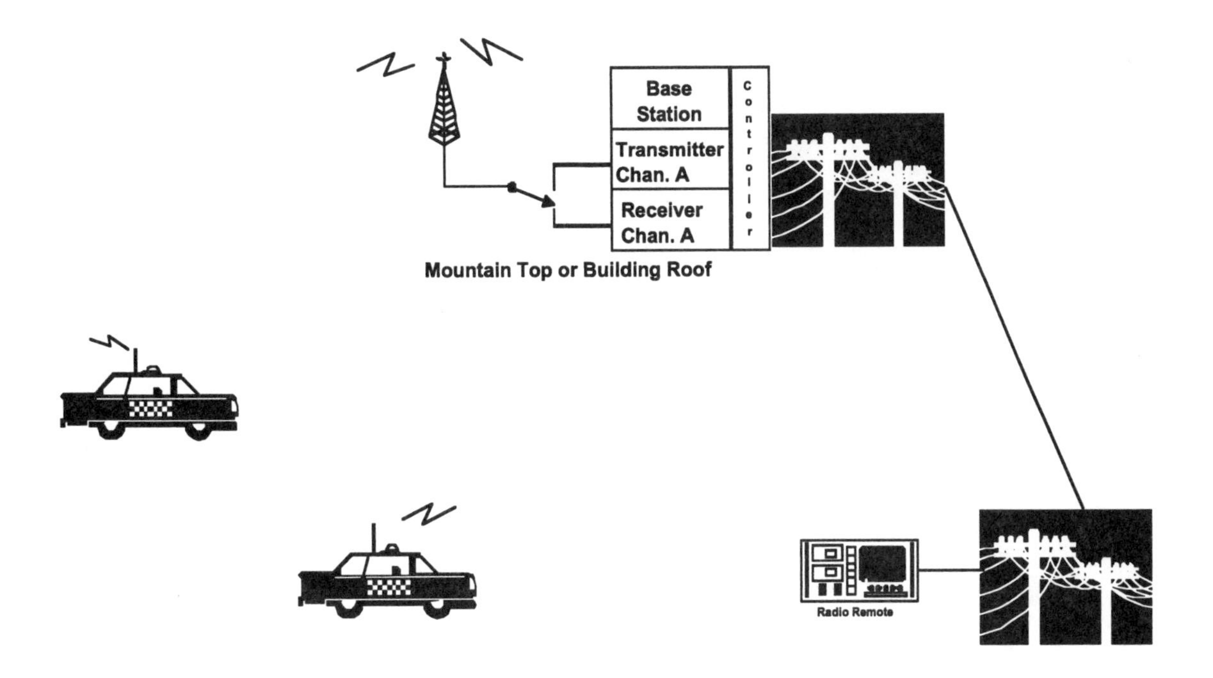

Figure A-2. *Remote Base Station*

Many systems operating on the low and VHF bands still function in this manner, but since telephone companies charge for their circuits by the quarter mile, many of the telephone circuits have been replaced with a radio or microwave links.

While business radio users were limited to this type of simplex operation in the low and VHF bands, some services, including public safety agencies, have been permitted to make use of repeater systems (see below). A repeater increases car-to-car range and, as a side benefit, enables the base station to be controlled using a radio link rather than telephone circuits.

HALF-DUPLEX RADIO SYSTEM

A half-duplex radio system differs from a simplex system only in that it employs two radio channels that have been "paired." Vehicles can only talk to the dispatcher—other cars cannot hear them. The FCC requires this type of system for such services as taxi dispatching (Figure A-3).

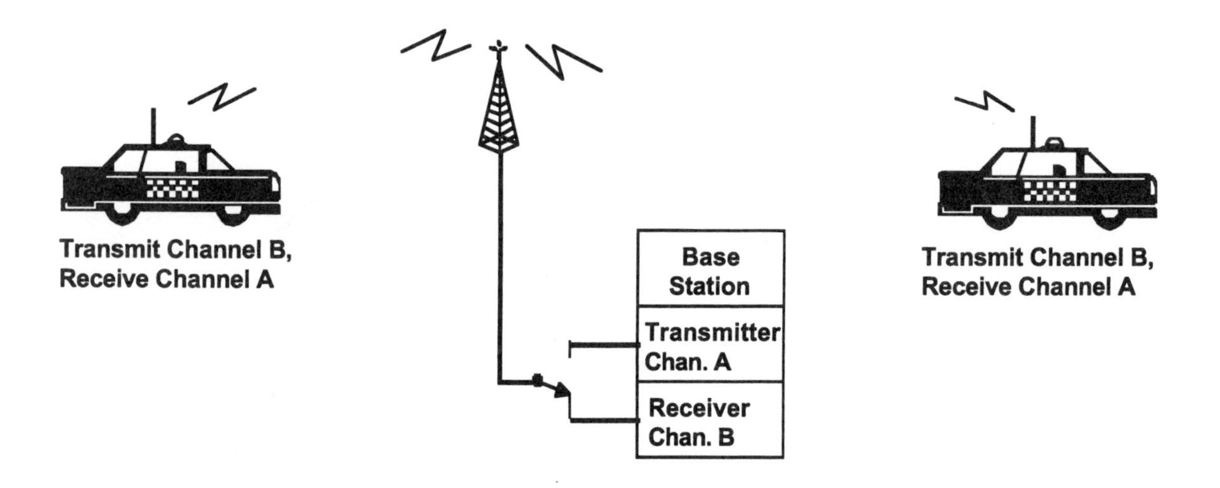

Figure A-3. *Half-Duplex Radio System*

The drawback to this system is that the vehicles cannot hear each other, and they tend to transmit on top of each other. This leads to confusion, since the dispatcher will not be able to understand *any* of the vehicles trying to communicate with the base. One solution used by many police agencies is to transmit a "beeping" tone on the base station channel whenever a mobile unit is transmitting. While this assures that only one vehicle transmits at a time, the beeping is annoying to those who must listen to it.

However, a half-duplex radio link is one of the most efficient means of sending and receiving data. This is because when sending packets of data, each packet must be acknowledged as having been correctly received. This acknowledgment is another packet of information that is sent from the receiving station back to the place of origin. In a simplex radio system, once the packet is received it must be checked, then the transmitter must be turned on, and the acknowledgment must be sent back over the same channel. With a half-duplex system, both transmitters can be on at the same time—the time lag between receiving a packet and the acknowledgment is decreased.

REPEATER RADIO SYSTEM

Expanding on the basic idea of a half-duplex system and turning it into one known as a "repeater" system was the next logical step. Using two antennas—or a special antenna mixer called a "duplexer," and special control circuitry—the received audio is fed out of the transmitter (Figure A-4).

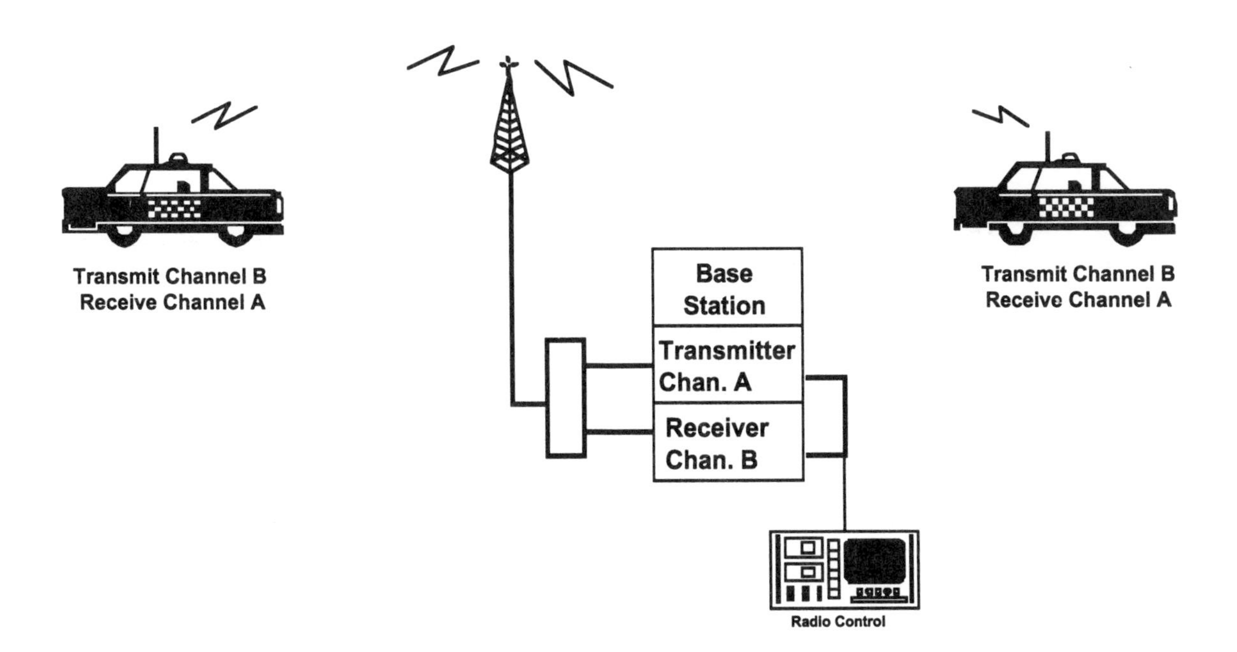

Figure A-4. *Repeater Radio System*

When a mobile unit transmits, the signal is received by the repeater and rebroadcast. The advantages are dramatic. First, every other vehicle within range of the repeater can hear the conversation. Second, if the repeater is located on top of a tall tower, building, or mountain, the system's range is increased. Another advantage is that the dispatch center can be connected to the repeater either by wireline or microwave—as in the remote base station example—or by a small, low-powered radio transmitter/receiver.

The entire system's mobile-to-mobile range is dramatically increased when repeaters are used. Instead of a mobile unit (or handheld) being able to talk with vehicles within the radius of only a few miles, the mobile unit talks through the base station. When the signal is relayed back from the base station, the effective coverage range of each mobile unit is equal to the coverage of the base station.

Matching Receivers to Transmitters

One of the most complex issues in installing a repeater system is matching the coverage area of the high-powered base station transmitter with the lower power and antenna gain of the mobile or handheld units. In an ideal world, anyone would be able to talk to the repeater transmitter from anywhere anyone else with a unit on that frequency could hear it. This is not the case, however. Radio engineers have become very creative over the years, using multiple receivers (as with the Mtel NWN system) or placing multiple repeaters in various locations, and tying them together with microwave or phone line circuits (techniques used by SMR and cellular systems).

Basic radio repeater technology is at the center of most modern radio systems. SMR systems—as well as today's cellular systems—are based on repeater technology.

SHARED RELAY SYSTEMS

Since there are not enough radio frequencies for all the users, many business radio users share a repeater with other users. Sometimes as many as ten or more companies will make use of the same UHF repeater. In such cases, each "fleet" is assigned its own designator and the units can only talk to other units that belong to the same company (fleet) (Figure A-5). In order to cut down the amount of radio traffic heard by all of these mobile stations and their respective base stations, a method called "Continuous Tone Coded Squelch System" (CTCSS) was developed.

Motorola calls its version of CTCSS "Private Line," or "PL," and this acronym has become the standard term for CTCSS. PL tones are transmitted along with the voice signal and are in the 88-Hz to 250-Hz range. When a PL tone is programmed into a radio, it prevents the radio's receiver speaker from turning on unless the signal heard by the radio is accompanied by the correct PL tone.

An important point to remember is that the tone activates the speaker when it is present. The radio receiver itself "hears" all the signals, or "traffic," on the channel, but the user does not hear any traffic that does not carry that unit's PL tone (because the speaker is not activated).

Sharing works effectively most of the time, but it does require a mobile user to wait when the channel is being used by one of the other fleets. Most of these systems are designed to temporarily disable the PL tone protection when the microphone is picked up so that a user can listen for any other traffic on the channel before speaking. If a second unit tries to communicate while the frequency is in use, there will be interference no matter what type of tone systems are being used. The problem is the same as that presented when one tries to start sending a fax on the same line on which the modem is checking e-mail.

The limitation of all these types of radio systems is that while they do permit communications to occur, each radio channel in a given area is limited to one conversation at a time. In public safety or dispatcher-controlled systems, the general rule is that one base station can handle up to about 100 mobile units on a single channel. In practice, this number is, more often than not, lowered to somewhere around 40—60 units per channel.

Thus, the City of San Jose Police Department needs to have more than a single radio channel for its use. In reality, it has somewhere in the order of 10 separate channels. Some of these are used for dispatching police cars (the city is divided into zones or sections and each section has its own channel), some for operations coordination, some for undercover and detective work, and one for administration and command activities.

San Jose Police vehicles and officers are all equipped with radios that can send and receive on all channels assigned to the department. Each radio has a frequency switch, similar to the push-buttons on a car radio. This switch is used to change the unit's frequency to the appropriate one. Most of the time, this system is very effective. However, if a car is switched to channel 3 while the dispatcher is trying to contact it on channel 4, the communications system breaks down.

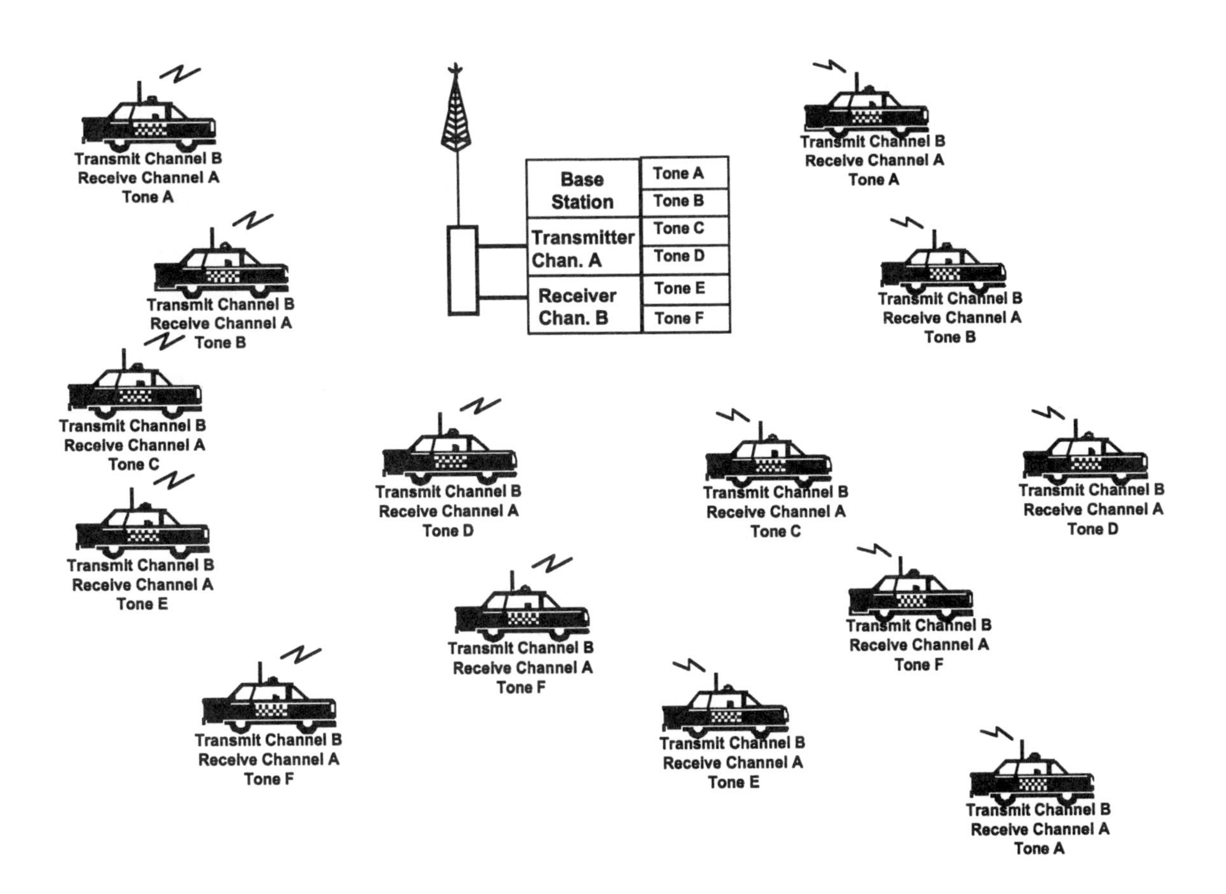

Figure A-5. *Shared Relay System*

SMR AND TRUNKED SYSTEMS

Over time, the FCC has granted new licenses for newer and more efficient use of the radio's channels.

Specialized Mobile Radio (SMR) systems make use of a technology referred to as "trunking." In this mode of operation, each base and mobile unit has a microprocessor controlled channel-changing system (Figure A-6). At rest (when no traffic is present), all units "listen" to the first channel. (There can be anywhere from 5 to 25 channels in these systems.) When a unit wants to make a call, the Push-To-Talk (PTT) switch is depressed, a digital signal is sent to the control system, and the next available channel is assigned for use by the requesting unit and his fleet.

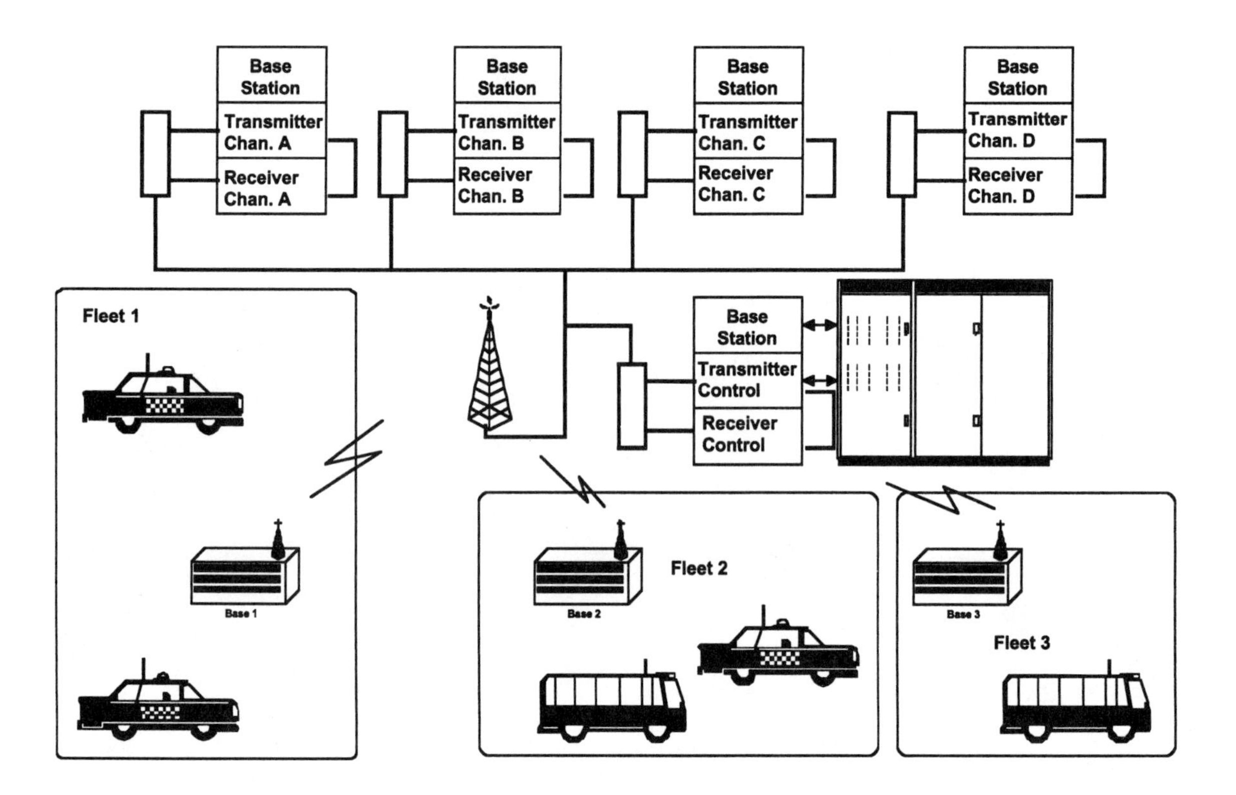

Figure A-6. SMR and Trunked Systems

A digital signal is then sent to all the units in that fleet and each unit is automatically tuned to the newly assigned frequency. After the transmission is complete, the units all return to the first, or main, channel and "listen" for the next request.

By making use of this technology and a multiple number of channels, more traffic between a larger number of units can be handled than in conventional single-channel radio systems.

SMR systems are generally placed into operation by companies that then provide the service of the computers and base station transmitters and receivers to others who want to make use of such a system. To the SMR radio users, there is little, if any, difference between the way they operate their radios for SMR and the operation of their previous systems. All of the control operations and the channel-changing are handled by computers and digital signals that ride along with the voice on the radio channels.

ONE-WAY RADIO PAGING SYSTEMS

For quite a while now, radio pagers have been an alternative to two-way communications. Instead of having a two-way radio installed in a vehicle, or having to carry a walkie-talkie, these small receivers clip easily to a belt or purse and they respond to a code that is sent out over a high-powered transmitter (Figure A-7).

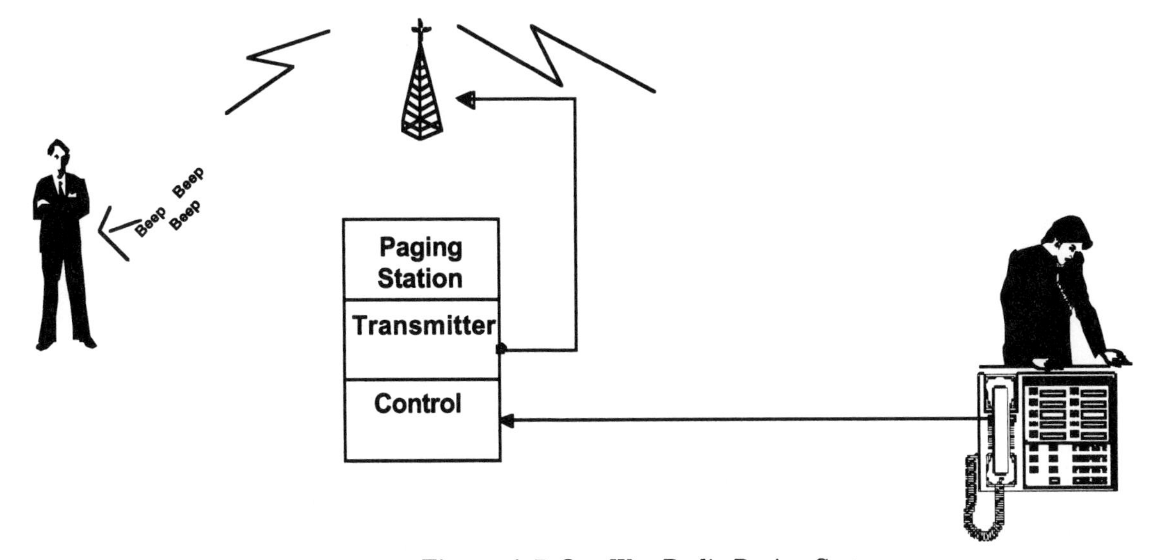

Figure A-7. *One-Way Radio Paging System*

Early paging systems used a simple two-tone paging tone to turn on an individual receiver and then a voice message was transmitted: "John call the office at once." "Dr. Smith, call the hospital emergency room, stat!" or some similar message.

Paging has become big business, and simple tone-plus-voice options have given way to many different types of paging tone formats. The greatest advance, so far, is the inclusion of data strings that are decoded and displayed on the pager as either numeric, alpha, or combination alphanumeric characters.

Paging range has increased. There are paging systems that cover metropolitan areas, entire regions, and at least three systems that cover the United States and parts of Mexico and Canada. Many of these paging systems will play an important role in the domain of handheld assistants over the course of the next few years.

CELLULAR SYSTEMS

Cellular is not the first type of wireless telephone system to be offered; it is the third. The first, called Mobile Telephone Service (MTS), was offered by a combination of Radio Common Carriers and phone companies and has been around since the late 1950s. These early systems were offered by the phone company, and they employed very few channels. There were few base stations, all of which were placed in high locations.

The next mobile phone service to be offered was IMTS (Improved Mobile Telephone Service). It provided a direct telephone dial type of service—users could make and receive telephone calls directly from their cars. Because the number of channels was limited, and because the systems usually consisted of only a few radio transmitters and receivers, users had to wait for system access and, sometimes, they could not get a channel to use for a call.

The cellular system concept was proposed to the FCC in the late 1940s, introduced to the FCC in 1953, and finally accepted as a licensable service in the early 1980s. The basic premise of cellular service is frequency reuse.

The System

Cellular telephone systems have been placed into operation around the world. At present, these systems are mostly using analog voice with digital data exchange to control the system. On a cellular system, many "miniature" radio sites are built, each of which has a very limited coverage area, and the frequencies are re-used in other sites (cells) in the same geographic area.

Each site is made up of many channels, ranging in number from only two or so, up to 24 or more, depending on the projected usage of the site. The combination of multiple channels per site and multiple sites per city permits the high volume of traffic on cellular systems today (Figure A-8).

As an example of the channel-handling capability, consider a cellular provider in the Philadelphia area (one of two) which has 332 channels allocated to the system in question. If a total of 50 sites are used to cover the city, and each site has an average of 24 channels, the system is capable of handling 1,200 simultaneous conversations, not just 332! The actual number of calls that can be handled by a given cellular system is probably higher still, but the point is made. By using the channels over and over again, many more users can be accommodated. As the number of users increases, the provider adds more cells–each closer to the others than before. Not only does the addition of a new cell increase the number of channels available in a given area, it also provides additional system capacity by increasing the number of times specific channels can be reused.

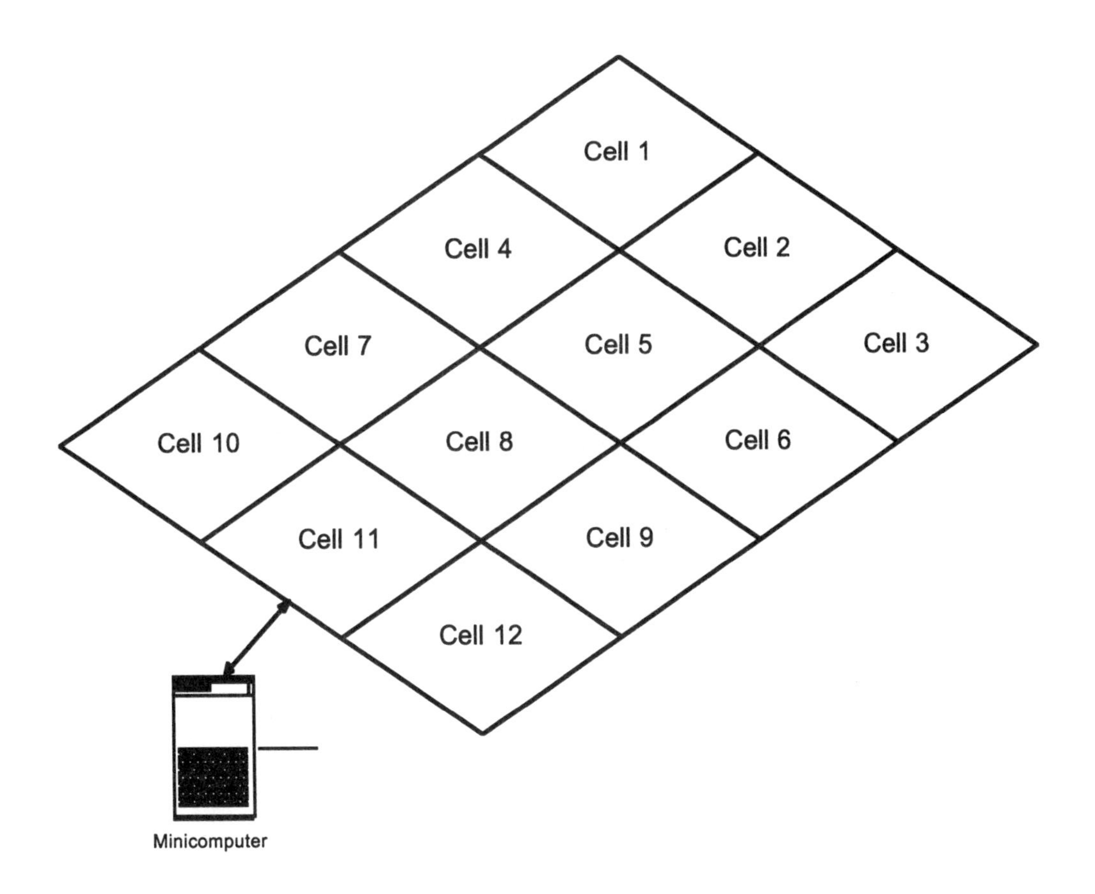

Figure A-8. *Cellular System*

Cellular (telephone) communications systems differ from standard two-way radio communications systems in several important ways. First, the object of cellular systems is to connect a single unit (customer) to the existing wired telephone network just as it would be connected in a home or office. Second, access is provided to the cellular system from the wired telephone network in the form of inbound calls. Next, the system is full-duplex, meaning that both parties can talk and listen simultaneously, just as in "normal" telephone usage. Finally, each user pays a monthly access charge as well as a per minute charge for the time used on the system.

Since the theory of cellular operation is to make use of a limited number of channels over and over again within the same geographic area, a master computer control system is used to coordinate the channels, sites, and users roaming around the area (Figure A-9).

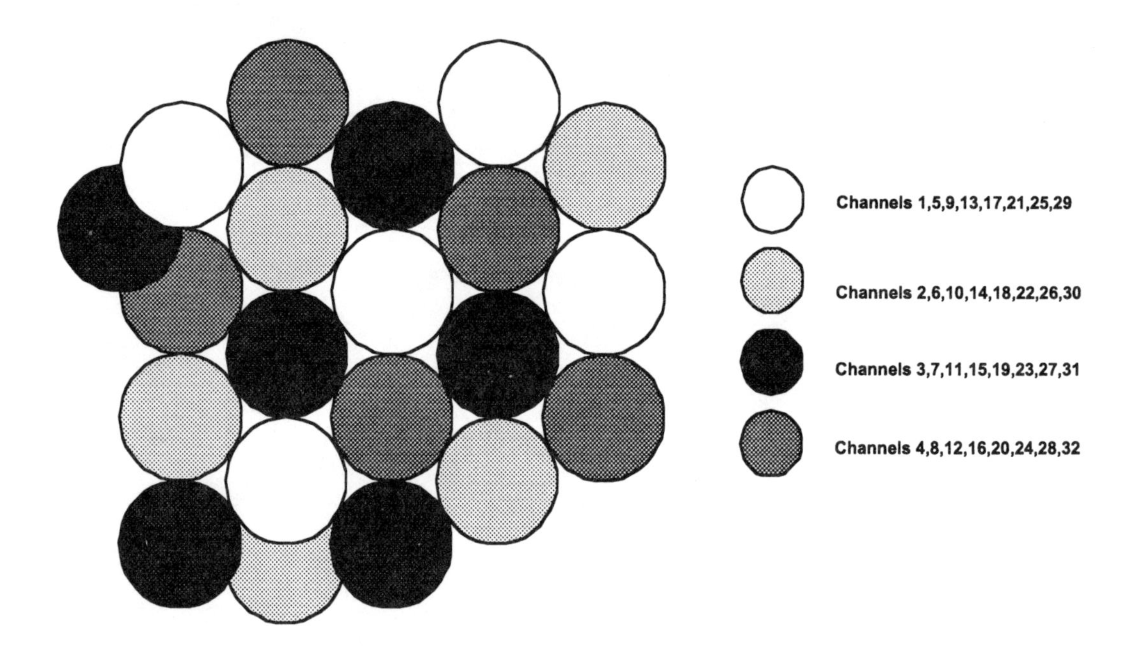

Figure A-9. *Cellular Telephone Frequency Reuse*

Each cellular phone (radio) is continuously "talking" to the cellular system. The digital information that is exchanged between each phone and the main system includes data indicating in which cell it is located, the output power of the unit (the central system can control the output power of mobile phones), and the phone number and Electronic Serial Number (ESN) of the phone.

Cellular Carriers

Each region of the country that has cellular coverage is served by two carriers. In the early days of cellular (early 1980s) these two service providers were easy to distinguish. One was the wireline carrier for the area (Regional Bell Operating Company (RBOC) or similar company) and one was a private company. The idea was to permit two competitors in every market, thus keeping pressure on service prices while providing a choice of system suppliers.

Today, the distinction between the two carriers has blurred to the point where it is rarely possible to distinguish between the carriers. RBOCs have taken partners; McCaw Cellular (one of the largest non-wireline carriers) has been purchased by AT&T (a wireline carrier); and some systems have become joint ventures between wireline and private sector companies. Regardless of who is running a service in any area, it is one of these two. Cellular phones are equipped with *all* of the cellular channels—you can change carriers by walking into a store and having the phone reprogrammed.

Cell Sites

Each cell site consists of a number of transmitters and receivers, antennas, and computer equipment to direct calls (Figure A-10). Each site is tied back to a master control point where the call is handed off to a wireline telephone service. The computer system necessary to coordinate all the channels and all the various sites is sophisticated, to say the least. Not only does it have to track each and every call, it needs to keep track of every phone that is operating on the system, and it must communicate with other systems as well.

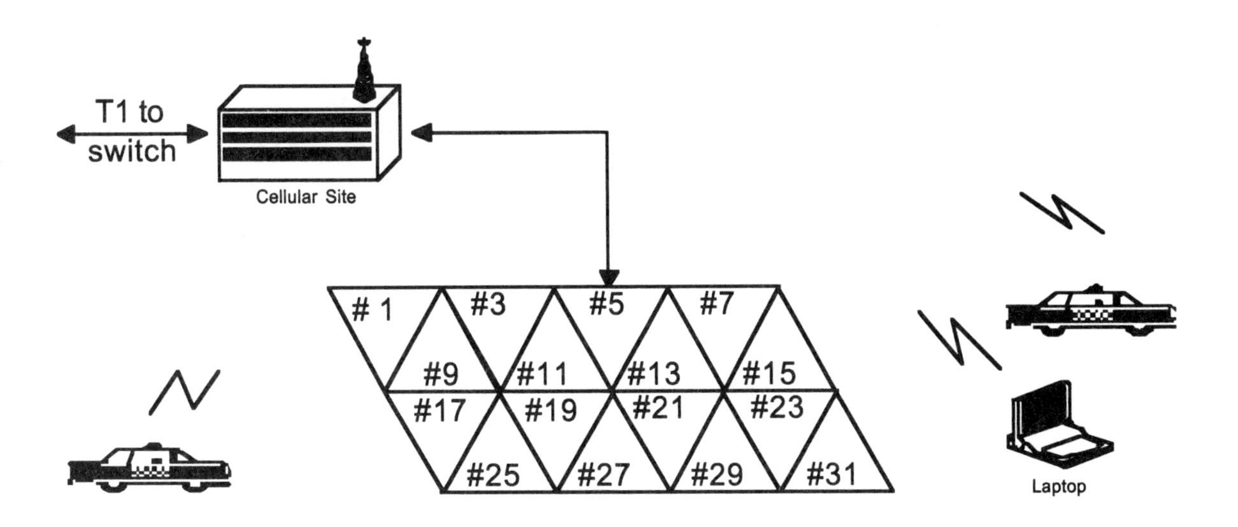

Figure A-10. *A Cell Site*

Because cellular systems are designed to be expandable, however, most service providers start with a minimum number of sites and add others as demand for the service increases. A system might start with only 15 or 20 sites located about 10 miles apart. As the number of users increases, sites are added by decreasing the spacing between sites and by adding sites where there are complaints about "dead spots."

Using Cellular Systems

Cellular users who use their units on a regular basis have experienced call drops, interrupted calls, lack of service availability, and poor quality reception. Still, these systems generally work well, providing a usable wireless voice communications link to more than eleven million U.S. users—and millions more on a world-wide basis.

Learning to use a cellular phone poses no real problem for most people. It is the same as using a standard telephone, with a few extra twists. However, what goes on in the system and in the cellular phone's control circuits to make it simple is extremely complex and is a real engineering feat.

The user merely dials a number, presses the send button, and then waits for the call to be connected. The system must allocate a channel within the nearest cell, switch that channel to a wireline, connect to the switch that will route the call, and keep track of the signal strength of the cellular phone during the entire conversation.

If the unit is moving, chances are it will pass from one cell to another; the system will have to switch the call to the new cell, select the best channel within that cell, and then send a command to change the frequency (channel) in the cellular phone. If the conversation is a normal voice call, this switching time (approximately 300 ms) is not noticed by either party and the conversation happens just as it would from a wired phone.

Difficulties Can Be Encountered

Because the system is not perfect, a user sometimes experiences a disconnect, or may even become connected with someone who is not the the caller's intended recipient. In some instances, more than one conversation can be heard at the same time. Many of these conditions exist because demand for cellular services has increased faster than the local company's ability to add or change cell sites. Some of the problems, however, can be traced directly to radio propagation.

Radio waves at 800 MHz travel only a short distance, usually line-of-sight. They can, and do, bounce off walls and other surfaces, and they are distorted and diminished by trees and other plants. Further, it is not possible to limit the distance radio waves will travel, nor is it always possible to predict how far from a cell site a given transmitter will be heard.

For all these reasons and more, it is possible to be receiving the same channel from two sites located a fair distance from each other and, subsequently, "hear" another conversation. Both conversations are heard with voice. If data were being sent over the same link, the reception would be garbled and, therefore, useless information.

At present, there are many different ways to send data over cellular phone circuits, but this is not easily accomplished with today's analog-based systems. Data transmission should become easier and less prone to error in the future, but, for the moment, anyone who wants to make use of the cellular system for data transmission, will need to have help getting set up and will need a lot of patience.

NETWORKS, ONE-WAY

Thousands of one-way paging systems are in use around the world today. Some of these belong to a specific business or hospital and are used to page (alert) personnel within a building or on a campus. Others are designed to provide local city-wide or area-wide one-way paging to users who subscribe to the service. The largest are those that have been designed to provide paging—and now messaging—services anywhere in the United States and, in some cases, in other countries.

Paging service providers run the gamut from a local entrepreneur providing service to a local area, to the Regional Bell Operating Companies (RBOCs), to specialized companies that provide large-scale paging operations. During the past few years, many smaller paging companies—especially those operating adjacent to a larger operator on the same radio channel—have been bought by larger companies.

Paging systems are licensed on the Low Band (30–50 MHz), High Band (150–174 MHz), UHF band (450–470 MHz), and the 800–900-MHz band. For the most part, paging consolidation has taken place in the High Band and in the 800–900-MHz band. High Band systems are still mostly populated by tone and voice, tone-only, and numeric paging systems, while 800-900-MHz systems (the newest) offer services well beyond simple message alerting and numeric call-back number displays.

It is in the 800–900-MHz band where most of the new paging and messaging activity will occur. The reasons are that the FCC has recently opened up new channels for paging and messaging systems in this chunk of spectrum, and because radio signals at these frequencies tend to better penetrate buildings and are less susceptible to foliage absorption that can decrease the range of a system.

Three systems—EMBARC, SkyTel, and MobileComm—are playing a major role in the conversion of an industry that has been providing alerting to subscribers to an industry that will play an important role in the wireless revolution.

EMBARC, SkyTel, and MobileComm

All three of these systems provide nationwide coverage and all cover the majority of the U.S. population. As with cellular systems, these messaging providers are constantly reviewing their own coverage areas as compared to the competition, adding radio transmitters to cover specific areas or adding other cities to their coverage.

The simplest way to understand how systems such as these begin to take shape is to step back and look at the components of a simple radio paging system.

The most basic paging system consists of an antenna located on the top of a building, and a radio transmitter. Connected to this is a paging console which sends a series of tones to the transmitter, turns the transmitter on, and sends the tones out over the air. Once the tones to alert the pager have been sent, the information intended for that pager is transmitted. Pagers evolved from tone-only (where the only indication was the beep or alert tone), to tone followed by voice, to numeric read-out pagers, to full alphanumeric messaging pagers.

With the increased sophistication of paging receivers, messaging formats have changed as well as the addressing capabilities of these devices. The first pagers alerted when they received a single tone; system expansion required the move to dual-tone. With the advent of dual-tone systems, it was possible to send messages to groups of pagers as well as to individual pagers. This was accomplished by making one tone common to all pagers within a group. Sending two tones (A and B) sets off a specific pager, but sending a single tone (A only) for a longer duration sets off all the pagers within that group.

As paging techniques became more sophisticated, and alerting formats shifted from tones to digital signaling techniques, it became possible to offer paging options well beyond individual and group alerting. Today it is possible for a user to receive individual messages, group-specific messages, and messages intended for a large number of different groups.

Building on the basic paging system concept, the next step was to add paging transmitters within the same city to gain wider coverage. The first such systems made use of sequential paging. With sequential paging, when a request to send a page is received by the service, it is sent to the first transmitter which sends the message. Seconds later, the same message is sent to a second transmitter and the message is sent over that one, and so on.

Obviously, this method of in-building coverage is not very efficient, since a single page occupies the channel not only during the first transmission, but also during the second, third, etc. The next technology introduced was a technique called "simulcast" where all the base stations in a system are turned on and transmit simultaneously (see Appendix B). Coverage overlaps of the various transmitters are carefully designed so that coverage is greatly improved.

Nationwide systems are based on tying a large number of metropolitan areas together and providing at least one central intelligent control center at the heart of the system. SkyTel, for example, makes use of a main computer center to process incoming paging and messaging requests, and then sends the information via satellite to each base station that is to be turned on (fired) (Figure A-11)

The intelligence built into the SkyTel, EMBARC, and MobileComm networks permits the messages to be sent to the entire system or to a specific portion of the system. In this way, these companies can offer service that is regional, national, or, in some cases, international in scope.

In addition to these three nationwide systems, there are presently a number of regional systems in operation, and the FCC recently allocated additional messaging channels for use on a nationwide basis. Further, some SMR carriers are offering paging capabilities over their existing two-way channels—a user can receive a message and then, using the same device, respond to it.

Advantage of one-way messaging systems is that the device worn or carried by the user can be very small, and it can be left on for months at a time without having to replace batteries. For example, a SkyWord pager, capable of receiving alphanumeric messages, operates for more than a month on a single AAA battery.

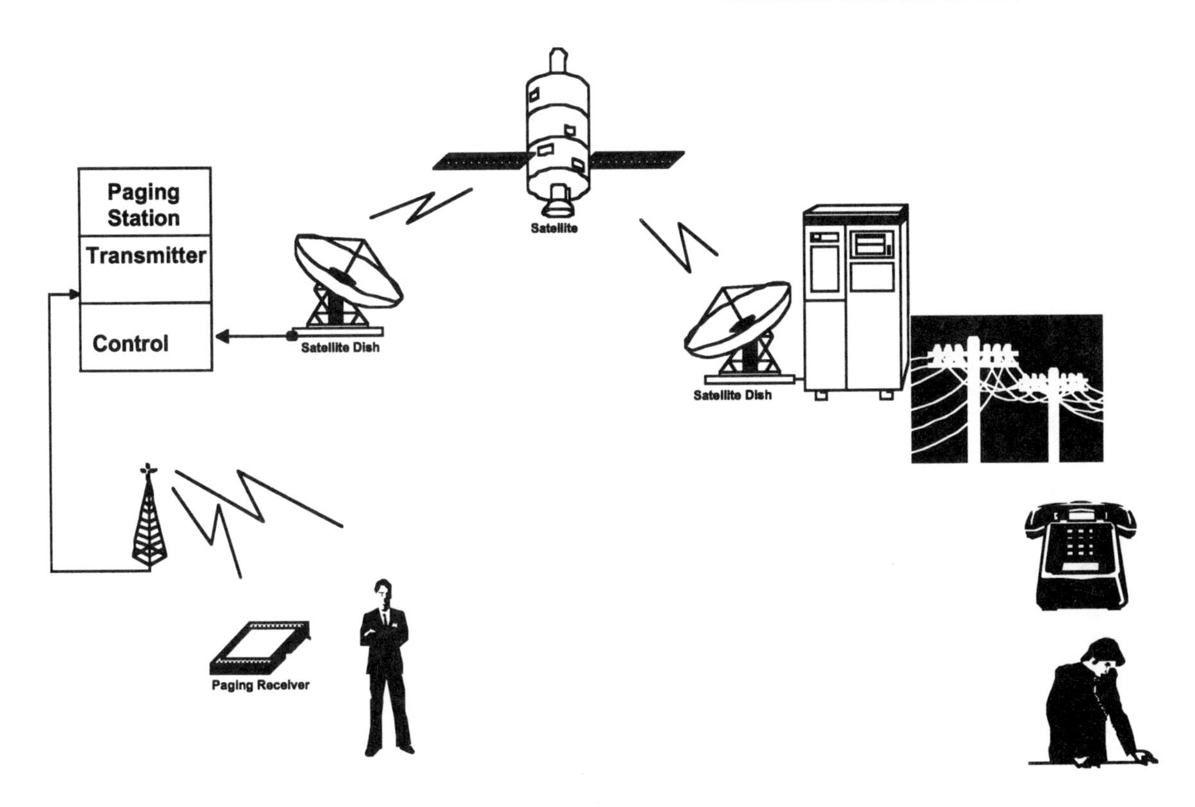

Figure A-11. *Nationwide One-Way Paging System (SkyTel)*

Future enhancements to one-way messaging products will include chip-level messaging receivers. One-way devices will become an integral part of handheld computing devices.

NETWORKS, TWO-WAY

Two-way wireless data systems must be designed to provide coverage from a base transmitter to a handheld receiver as well as to complete the path back from the handheld device to system receiver. The key to the success of these systems is that they *must* provide extensive coverage so that users do not have to think about whether or not they are within range to access a base station—the systems must provide coverage in most places within the United States.

One reason that computer companies and users alike seem to believe that CDPD (Cellular Digital Packet Data) will work so well is that the voice component of the system already works well in most metropolitan areas—even to and from small, low-powered handheld cellular phones.

However, no one system offers 100-percent wireless coverage of the United States. Even when the satellite systems are deployed later this decade, there will never be a system that provides the coverage needed for truly ubiquitous operation. Instead, the networks of today have been designed to cover as much of the large population centers as possible, with expanding coverage to meet the demands of the end users.

At present, the three ways to send and receive data on a nationwide basis are over analog cellular (voice), ARDIS, and RAM Mobile Data systems. By the middle of 1995, two more wireless services will be in operation: Mtel's Nationwide Wireless Network (NWN) and CDPD. A few years into the future, the number of networks capable of sending and receiving data could climb to eight or ten systems (including PCS).

Such two-way systems typically fall into three configurations. In a *trunked* system, the command channel switches the units to a free channel and establishes the communications link. In a *mesh* network, each node can talk to any other node it can hear. Traffic is not always routed over the same set of nodes. A *star* network consists of a central controller that routes the message to the proper radio transmitter.

It is also important to understand that some of these systems were originally designed to provide coverage to and from radios rather than to and from a low-powered handheld device that might be located inside a building. For example, most SMR systems were designed to provide communications to mobile radios installed in cars. Most such vehicle radios operate at power levels of between 20 and 60 watts. The first generation of cellular telephone systems were designed to provide communications to and from cellular telephones but with a transmit power level of only 3 watts.

Today, the criterion for cellular systems is to provide coverage to handheld phones operating at a maximum power level of only 600 milliwatts–a far cry from the 3-watt car phones, and even further from SMR radios. RAM and ARDIS, on the other hand, were designed to provide in-building coverage to handheld devices from their inception. Power levels of the first ARDIS and RAM wireless units are in the 2-watt range, but the next generation of products will also be in the under-1-watt range.

Base Station Placement

Since SMR systems were designed primarily for dispatch functions and to work with high-power mobile units, they generally made use of a single radio site located well above the terrain it is to cover—usually on a high tower, on top of a tall building, or on a high mountain. This single-site approach provides good coverage to high-powered mobile radios, but it does not provide adequate coverage for portable radios or in-building communications. Therefore, SMR operators who want to offer service to handheld units must redesign their systems, adding transmitter and receiver sites and installing costly command and control equipment to optimize their systems for the new users they hope to attract.

Likewise, cellular service providers that originally built cell sites on 10 or even 15 mile centers—primarily to cover major highway arteries within a city—have had to add cell sites, making those sites closer together. In some cases, sites have even been installed within a building complex to provide the required coverage to handheld phones. Since developing a cell site is an expensive proposition, cellular providers tend to add cells only as demand requires. As an illustration of the competition factor, one has only to listen to cellular providers' ads on local radio stations. When GTE MobileNet installed a cell site that provided coverage through the Caldecott tunnel (a major commute route to and from San Francisco) its ads extolled this "great" feat of engineering. Once its competitor (Cellular One) also provided coverage inside the tunnel, these ads from GTE MobileNet ceased.

Enter ARDIS and RAM

When ARDIS first installed its system for the use of the IBM field service force, it was clear that the points of access had to be from within buildings, usually from within the bowels of the buildings where copier machines and computers are located. (Who ever heard of a copier or mainframe being assigned a window office?) From its inception, the ARDIS network was designed to work from within buildings. Care was taken to locate radio transmitters and receivers in close proximity to business centers.

One of the finer points of radio communications engineering is that in order to cover a wide area with a single transmitter, that unit must be located as high above average terrain as possible, but building penetration will be sacrificed for extended coverage. For this reason, ARDIS chose its radio sites on lower building roof-tops that "look" into other buildings. Here again, it is easy to get caught in a radio communications trap. To provide radio coverage within a twenty-story building, the antenna should *not* be located on top of that building. Rather, it should be placed on an adjacent ten-story building. Signals are directed outward from the antenna, not down below it. In communications parlance, the area directly below an antenna is referred to as the "cone of silence."

When the RAM system was in its planning stages, it also designed its system to provide maximum coverage inside buildings. RAM followed the same guidelines as ARDIS: More, lower antennas are better than fewer, higher antennas. There is still a question about which system provides the best in-building penetration for wireless data users, but like their cellular competitors, for the most part, this is a moot issue. If ARDIS has better coverage inside a major building today, users can be assured that RAM will add another base station and "catch up" to ARDIS, and vice versa. RAM and ARDIS have a real advantage in the deployment of radio transmitter/receiver sites over cellular operators because even the most expensive RAM site costs a fraction of a new fully-functional cellular site.

System Specifics

The ARDIS nationwide wireless network is built around a single radio channel that is licensed by ARDIS on a nationwide basis. In areas where traffic demand exceeds the capacity of this single channel (or is likely to exceed it), ARDIS has acquired additional channels. The first ARDIS field units were all single-channel radios with a 4,800-baud data rate capacity.

Motorola InfoTac units that are available for the ARDIS system today are multiple-channel units with a dual data-rate of 4,800 and 19,200 baud. When in an area covered by a single channel, the system operates at 4,800 baud. Where multiple channels are available, the first choice for the device will be a 19,200-baud channel. The external InfoTac operates for four to six hours on a single battery charge (the ARDIS network does not yet have battery-saving features enabled), and the transmitter operates at a power level of 2 watts.

The ARDIS Backbone

The ARDIS system makes use of multiple radio transmitters and receivers in each city in which it provides coverage. The number of stations is increasing on a weekly basis, so specific numbers are not as meaningful as they might be. However, by mid-1993, ARDIS had a total of more than 1,300 base stations deployed nationwide. These base stations are all connected to one of 35 radio node controllers which,

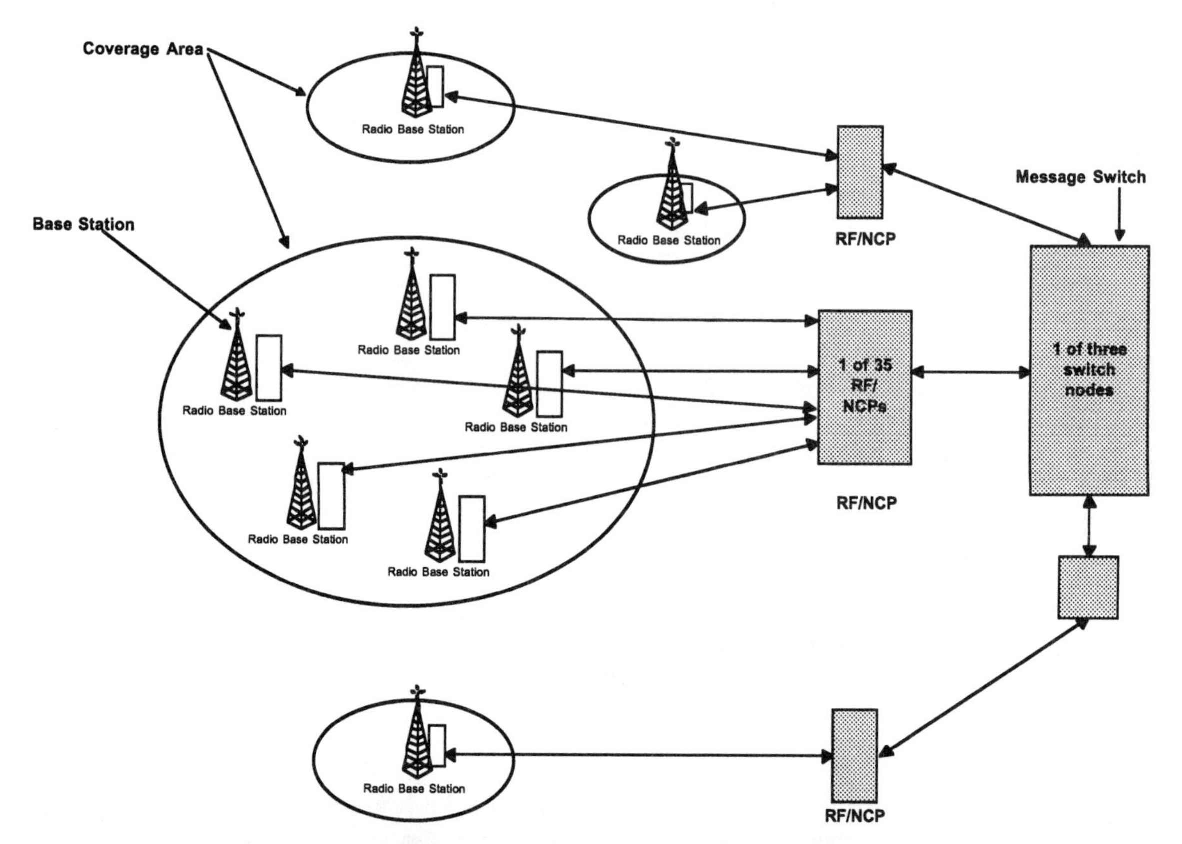

Figure A-12. The ARDIS Wireless Data Network

in turn, are connected to three message switches (only one of which is on-line at a time; the other two are fully redundant back-up systems). The system (Figure A-12) covers more than 8,000 cities comprising 400 geographic areas.

Each node controller is connected to the three switches with several different wired and microwave links. Each of the 35 node controllers is inside a "hardened" site provided by phone companies. The entire system is run by a command and control center so sophisticated that an operator sitting at the main console can "zoom" into the network and watch a data transmission take place between a single user and a base station.

RAM Mobile Data

The RAM Mobile Data system in the United States (Figure A-13) is modeled after systems already in operation in Sweden and the UK, and those being deployed in other European and Pacific Rim countries. It is basically an adaptation of an SMR system in that it makes use of between 10 and 30 radio channels per geographic area.

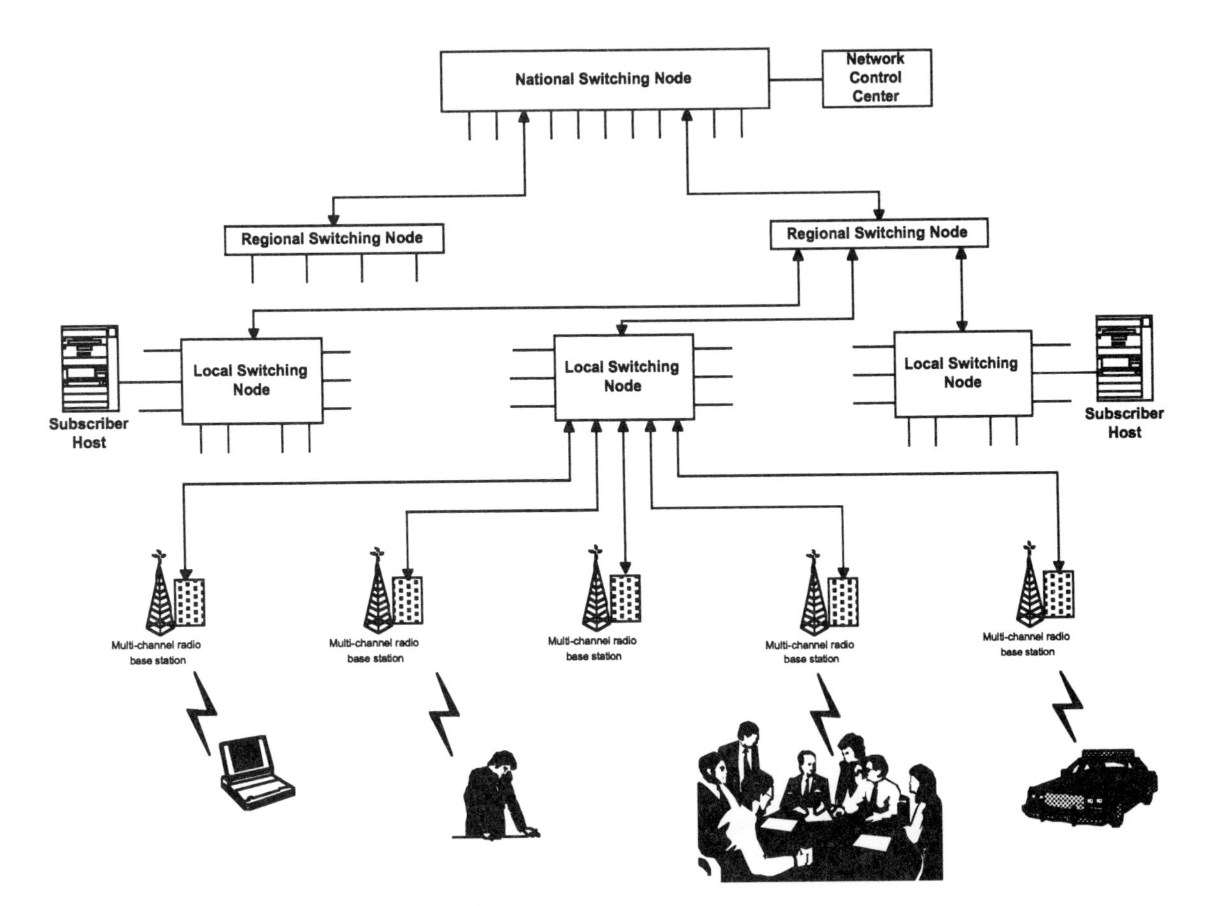

Figure A-13. The RAM Mobile Data Network

As with ARDIS, it was developed to provide coverage to low-powered handheld devices inside buildings. Base station locations are chosen for both high-level wide-area coverage and low-level, in-building penetration. Each base station is controlled by a local switching system which, in turn, is controlled by a regional switching arrangement. The regional switching node, in turn, is controlled by the National Switching Node.

The network control system is central to the network—it controls all aspects of network operation and data-flow. Companies connecting to the system for company-wide use do so at the local switching-node level, and they have access to the entire nationwide system.

When a user requests access to the network, a unique ID code is sent to the system and the user is automatically routed to the proper point in the network. If a user of the RAM system is to be connected to his or her company's mainframe or host computer, the connection is made automatically. If the user is a RadioMail customer, the connection is automatically made to the RadioMail gateway and the user can communicate directly with the RadioMail host.

The RAM system, which completed phase one of its United States build-out in 1993, consists of 800 base stations covering 100 Metropolitan Statistic Areas, or 90 percent of the U.S. population.

Common Features

Both RAM and ARDIS provide for seamless nationwide roaming, and the user receives a single bill at the end of the month. Unlike "roaming" on cellular networks, there are no roaming charges. A message sent from Los Angeles to New York costs the same as a message sent within a city.

Perhaps the most user-friendly aspect of the ARDIS and RAM systems is that users do not have to notify anyone of where they are. They can interrupt their network access in Los Angeles, board a plane, and reestablish network contact in New York without missing a single message. The networks simply reestablish the connection and continue from where they left off. When sitting in an airport waiting for an airplane, users do not have to be concerned about "ending" a session or signing off the network, they simply turn off the computer and modem and walk onto the plane.

General Comments

RAM and ARDIS are the *only* wireless two-way networks in operation today that were designed from the ground up for wireless data transmissions. They have been optimized for data, and they are fully deployed. The technologies have been fully tested and are functioning well. While one may have the advantage in terms of coverage or information access at a given point in time, they are fierce competitors and the advantage will continue to shift back and forth. For all intents and purposes, RAM and ARDIS offer the same level of service, and they are both ready for prime time.

One final observation. Both RAM and ARDIS contend that their systems can handle a large number of users. The math to prove this point is available and applies equally to both networks. For the immediate future, there is no danger of either system running out of capacity. A fundamental difference in the systems is that ARDIS operates on fewer channels, but acquires additional channels in areas of peak demand. RAM has already invested heavily in acquiring its channels on a nationwide basis.

DATA OVER CELLULAR

Many computer and communications industries people believe that the cellular system is the total answer for providing wireless data communications. To justify this choice, they point to the large number of base stations already deployed (more than 8,000), and the claimed superior coverage of cellular systems.

There is no winner in this debate. Personal preference and actual experience are most important. However, one point in favor of RAM and ARDIS is that they have been designed to be wireless digital data highways. Cellular systems were designed to provide analog voice service to users.

There are two ways to send and receive data over cellular networks. The first is to make use of voice channels and "session connections." The second is with a new technology referred to as CDPD (Cellular Digital Packet Data).

The cellular system certainly has many more base stations than does RAM and ARDIS combined, but its in-building coverage, while adequate at present, needs to improve if cellular is to be considered a major wireless data competitor.

Using Cellular Systems for Data

The simplest way to make use of existing telephone networks for data transmission is to buy a special cellular modem and connect a phone to one side of it, and a computer to the other side. Using standard modem software, it is then possible to dial up a network, or call a PC, and establish a connection. This connection is called a "session"; the computer and software expect the modem to stay connected to the network during the entire session.

If a caller changes location, or if another user initiates a call, the connection can be broken. Any information in the process of being sent or received will, most likely, be lost and will have to be retransmitted once the connection is reestablished.

New Technologies

New technologies have been introduced to minimize the problems associated with sending and receiving data over analog cellular systems. Some systems are providing a modem pool at their sites for better connections to the wired telephone network. Other companies, such as Microcom and AT&T, are developing new error-correction protocols.

Still, the use of analog cellular voice circuits for sending and receiving data is marginally acceptable at best. However, for the occasional user, it may provide enough of the basics to be worthwhile as an alternative to dedicated wireless networks.

Analog Cellular Data

The main advantage to sending and receiving data over a standard analog cellular system is that doing so most closely approximates the use of a standard wireline modem, and generally requires no modification to users' applications.

CELLULAR DIGITAL PACKET DATA

The cellular industry has embraced wireless data in a big way. A digital packet technique was invented by IBM and given to the cellular industry. It was originally called "CelluPlan," and it has become known as "CDPD." CDPD uses a number of digital techniques to send and receive data over idle cellular voice channels. These techniques include having the data "hop" from vacant channel to vacant channel (frequency hopping, Figure A-14). Theoretically, it permits cellular system operators to offer digital data on their systems without impacting voice traffic.

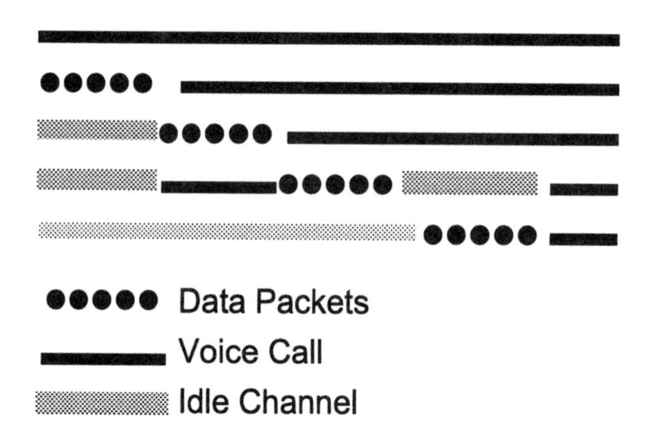

Figure A-14. Frequency Hopping Used in CDPD

CDPD is being rolled out over many of the cellular systems in the United States. At Comdex/Fall in Las Vegas in November of 1993, McCaw cellular demonstrated CDPD using both the hopping technique and the use of dedicated channels.

Of all the data schemes proposed for wireless communications, CDPD is the most complex and the most ambitious. First, a truly nationwide cellular network does not exist today. Analog voice users must often pay a premium, or "roaming" fee, for use of a system other than their home system. Further, the cost per minute when in the roaming mode is normally two to three times what a local subscriber pays.

Another point to consider when looking at the technology differences between dedicated wireless data networks and CDPD, is that the networks are just that: networks with intelligent controllers that control the network and route the traffic. CDPD is, in reality, a data pipeline over which information can be sent. But the traffic has to be directed by some external method (the end user "dialing" a connection, or a "smart" switch located somewhere along the network).

In theory, CDPD appears to many within the computer industry as the "winner" in the wireless data network wars. This perception has been helped by the marketing efforts of the cellular carriers and the fact that the computer vendors do not want to be required to offer a series of options. Therefore, vendors hope that CDPD will become pervasive enough to become the prime data transport system.

The reality is that at this time CDPD is still an unproven technology, without an "owner" that can control its roll-out and nationwide implementation. CDPD is being deployed by some cellular carriers, but not all. There does not appear to be a concerted effort to establish a single user, single bill system or to handle differences in specifications and operations from one system to another.

CDPD Conclusions

CDPD will play an important role in wireless data. However, it is not yet clear if this role will be mainly for point-to-point communications links needed to monitor vending machines, or systems that will truly empower users to become mobile and remain connected to their data. Time will tell. For 1994, CDPD technology should be considered as in its "pilot," or beta test stage of its deployment.

NATIONWIDE WIRELESS NETWORK (NWN)

The NWN system presently under construction by Mtel is based on a derivative of its existing SkyTel nationwide network. NWN was awarded a Pioneer's Preference License by the FCC for a 900-MHz frequency. A Pioneer's Preference License is issued only when the technology to be employed is determined to be distinctive and different enough that its implementation advances the state of the art.

The characteristics of the outbound channel, the channel from the network to the handheld device, are almost identical to those used in the SkyPage system. Both use many high-powered base station transmitters in each metropolitan area, and both employ simulcast techniques. All the transmitters are controlled by the network through satellites, but because the radio channel is wider than the SkyTel outbound channel (50 KHz), the outbound messaging speed is faster (24,000 bps) (Figure A-15).

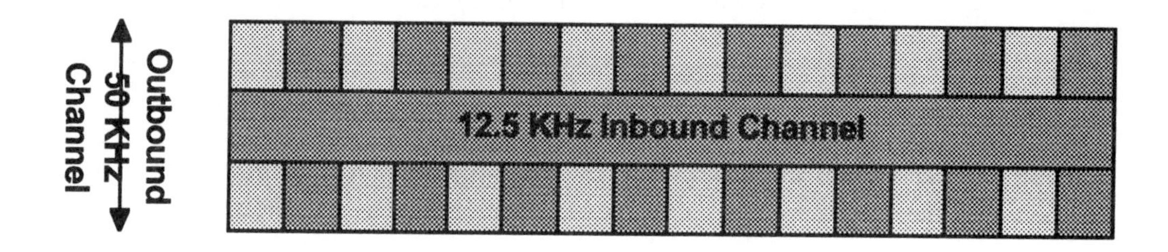

Figure A-15. *NWN Channel Usage*

The inbound channel architecture sets the NWN system apart. Instead of each high-powered transmitter being paired with a single receiver, each transmitter's coverage area contains multiple receivers at various sites. NWN realized that for handheld devices to be effective for two-way messaging, the system would have to be designed in such a way that the low-powered transmitters in the handheld devices can be "heard" by the system. While most other systems deploy a single transmitter/receiver combination at each site, the NWN approach is to install fewer, more powerful transmitters and surround them with more (and less expensive) receiver-only systems (Figure A-16), thus matching the system's receiving range with that of its transmission range. The receivers are tied back into the network with telephone circuits.

The result should be a system that provides extremely good coverage—even inside buildings. After the system is deployed, if NWN finds a receiver coverage "hole," it can deploy another receiver rather than having to install a compete transmitter and receiver combination to provide the coverage.

Because the handheld-to-network channel has less bandwidth (12.5 KHz) than the outbound channel (50 KHz), the inbound messaging speed will be 9,600 baud. Since each transmitter sends all of the outbound data in a digital format, packets will be "stacked" on the outbound channel, just as they are in any other packetized system. However, since a handheld device will only be able to access one or two of the system's receivers, the system will be able to support many handheld units.

NWN's vision is to provide pocket-sized devices that function as "acknowledgment" pagers. That is, the smallest form factor devices will not have to be equipped with a keyboard or other computer-related functions. Instead, they will be about the size and weight as present-day pagers, but they will include a transmitter and several pre-programmed message buttons. After receiving a message, the user would push one of the buttons to send the pre-programmed response–perhaps a message that says, "I have received your e-mail message and will respond to it as soon as possible."

NWN also expects to offer complete two-way messaging services, but it feels that a large potential for acknowledgment paging also exists.

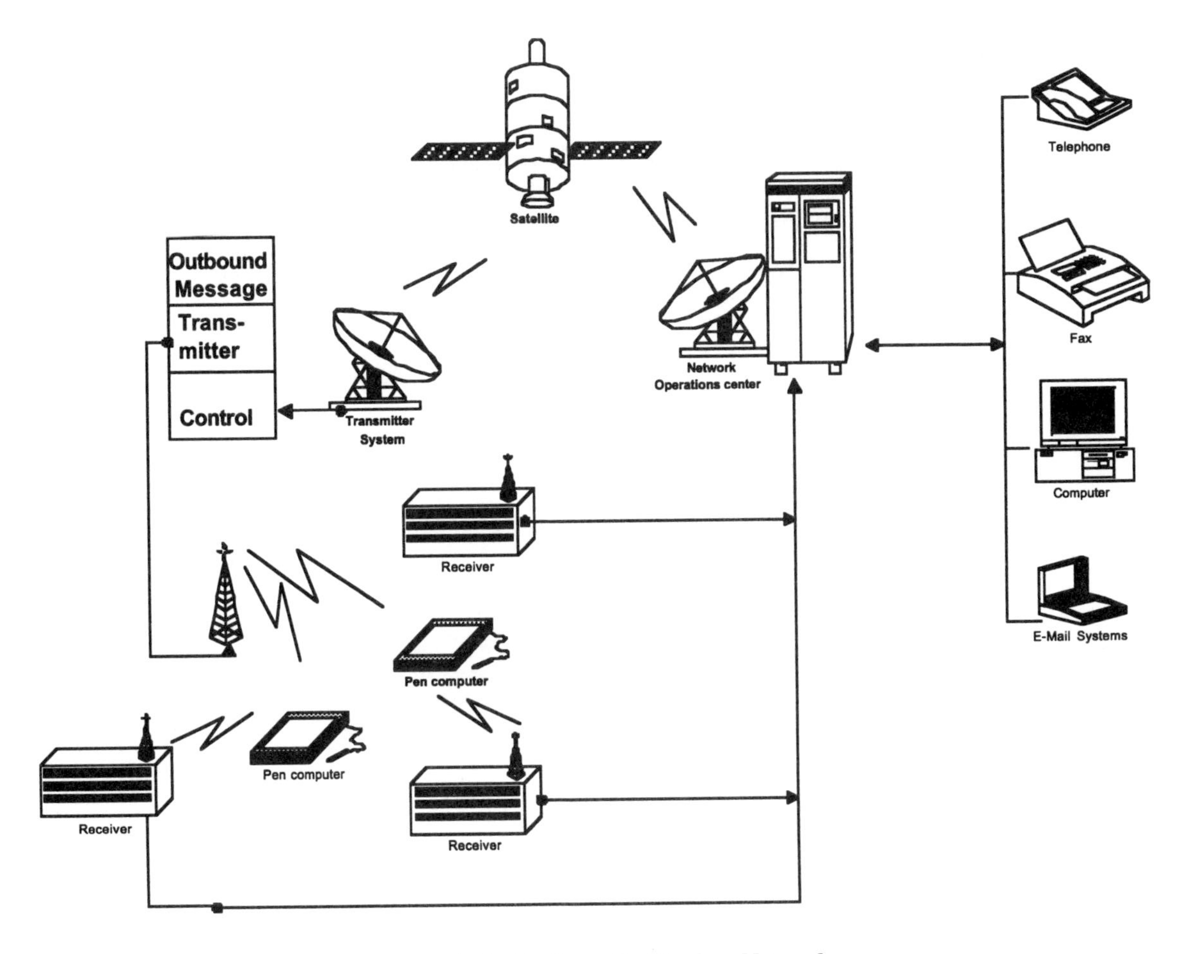

Figure A-16. *Nationwide Wireless Network*

METRICOM'S MESH NETWORK

The Metricom system concept more closely resembles the PCS model than that of any of the other wireless connectivity service providers. At present, the Metricom system occupies a portion of the 902–928-MHz unlicensed band, which limits the power of its transmitters but provides some interesting opportunities.

The system is based on spread-spectrum technology, so the data rate is higher than that of any of the other systems. Speeds of up to 56 Kbps or higher are achievable using the Metricom technology.

As designed, the system is applicable for in-building LANs as well as campus-wide and metropolitan areas. It is unique because when two handheld units are in range of each other, they can communicate directly. Since they are not using the network backbone, this type of communication is not subject to network usage charges. When the units are not near each other, they will "look" for the closest node and establish communications through it. Once a handheld is connected to a node, the node will route the transmission from node to node until it reaches its final destination.

Because the system is based on a "mesh" design (Figure A-17), each system node is capable of sending and receiving from any other node in the network that it can "see." Thus, traffic from one point to another will not always take the same path, it will take the best route at that time, sometimes sending the return information back over an entirely different path.

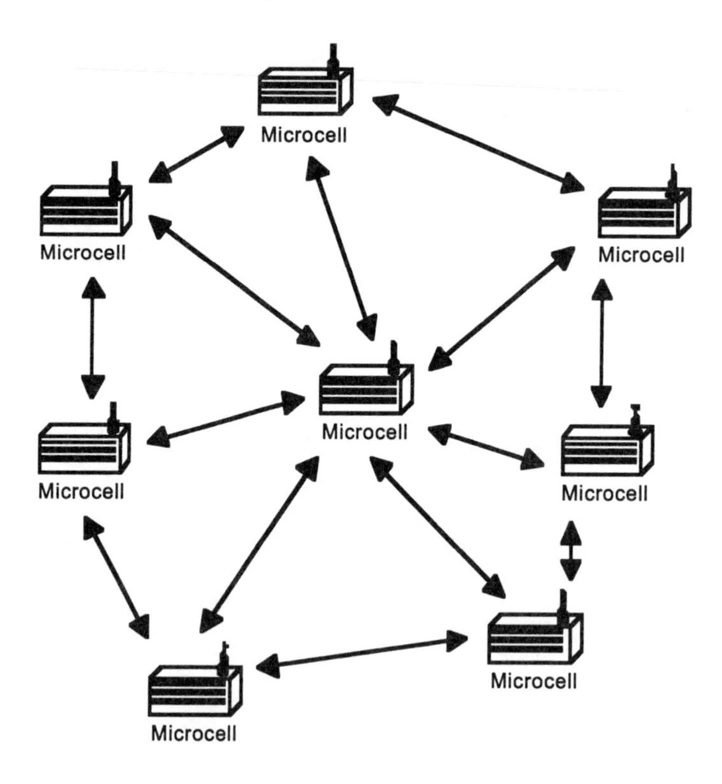

Figure A-17. *Metricom's Mesh Network*

Because of this unique architecture, the Metricom system can be best characterized as a cross between a wide area and a local area network. It is conceivable that a business could install a Metricom system on its own campus for internal use and have access to the entire metropolitan area through Metricom's own network deployment. Another feature of the system is that Metricom has elected to offer monthly unlimited access charges based on the speed of the data transmission, not on the amount of data transmitted. If a user elect to make use of a 9,600-baud connection, the monthly fee would be lower than if that user opted for a 19,600-baud connection or a high-speed link of up to 56 Kbps.

Another advantage offered by Metricom is that it deploys its system so that each "node" is housed in a self-contained shoe-box-sized housing that can be installed almost anywhere that AC or solar power is available. Since each node is inexpensive (less than $1,000)—even though more nodes will be required to cover a given area—the total system cost will be lower than that of any of the other metropolitan coverage systems.

PERSONAL COMMUNICATIONS SERVICES

The FCC has authorized up to seven different sets of frequencies for PCS. Two of these are 30-MHz blocks, one is 20 MHz, and the remaining four are 10 MHz. It is possible that all seven will be licensed in major metropolitan areas. What is not clear at press time is how many of these systems will offer voice-only, data-only, or a combination of voice and data. Since the technologies employed from the outset will be digital, all of the systems should be fully capable of offering both services.

PCS is intended to offer voice service that will provide a single phone number for a user. When a user is at home, the system will connect to that user's local PCS unit, and there will be no charge for network services. When the user leaves home, he or she will access the PCS network and pay for airtime. From the office, phone will connect to the office PBX system and, once again, no service charges will be incurred. The PCS data vision, the type of service to be offered, and the expected charges are not yet well-defined. By the end of 1994, the licensees and their ideas for the types of services that will be offered should be known.

For additional information regarding PCS, see the reference listings in Appendix C.

Appendix B

Technologies Within Technologies

WHO RULES THE (RADIO) WAVES?

As previously stated, there is a limited amount of usable radio spectrum. Over the years, technology has pushed the usable spectrum upper limit higher, and the available spectrum is being used more efficiently. But the fact remains that even when technology enables use of all the available spectrum, it will never be enough to meet the demand.

Some of the espoused visions depict everyone running around the world with a personal communicator, sending and receiving full motion video, voice, and information. The reality is that there is not enough spectrum available to permit such usage. Those offering such visions are not those in the labs who have to work within the constraints of physics and frequency allocations. They are, rather, visionaries who have not reconciled their visions with reality.

The first impression given from the frequency charts is that a great deal of spectrum is available for personal communications. After all, the total usable radio spectrum is more than 2 GHz; surely that is enough to support these visions!

However, the world body that "controls" the spectrum has been wrestling with demand versus allocations for a number of years. Each time this organization convenes, it must settle on compromises, and it must permit some services to share spectrum.

Spectrum Pecking Order

The spectrum allocation process begins with a world body called the "World Administrative Radio Conference" (WARC). The entire organization meets every few years to review existing use of the spectrum and to make new allocations as required, and as available. WARC's goal is to try to retain some order to a process that could create chaos if not controlled.

It is a difficult and thankless task. No country gets everything it wants, and sometimes a plan to coordinate spectrum usage cannot be agreed upon. The most recent session of WARC was held in Spain in 1991 to set up allocations for Personal Communications Services (PCS), satellite communications services, and others. The chunk of spectrum that the United States has chosen to use for PCS is not the same as that which will be used for the rest of the world. This means that U.S. systems will not be compatible with those of the rest of the world.

If WARC is responsible for spectrum allocations, how can the U.S. fly in the face of this group and do its own thing? The answer is that the U.S. did not do anything that is not in compliance with the WARC treaty. WARC does not allocate spectrum other than by reserving it for specific types of services either on a worldwide basis (for satellite communications, for example) or by region. The U.S. PCS falls within the general allocations structure of the treaty, but it does not fall within the specific frequencies being used elsewhere.

This type of discrepancy is not new to WARC. In the higher frequencies, there are more deviations in the allocations, making it more difficult to coordinate frequencies. At lower frequencies such as 2 to 30 MHz—where radio waves can easily travel around the world—coordination for specific services on a worldwide basis is essential. Broadcast stations such as Radio Moscow and Voice of America are found in the lower frequencies. The portions of these bands that are allocated to amateur radio worldwide are basically the same for every country, as well. There are a few exceptions to this but, for the most part, lower bands are allocated on a worldwide basis.

However, in the portions of the spectrum that are considered line-of-sight communications—higher frequencies, above about 100 MHz—it is possible to find countries using this spectrum in different ways. There are many examples of this. The RAM Mobile Data Systems in Sweden, England, The Netherlands, and other European countries operates in the 400-MHz band. In the United States, however, the RAM system operates in the 900-MHz band. By the time RAM applied for frequencies in the U.S., the 400-MHz band was already assigned to other users.

BEYOND WARC

Once allocations are made by WARC, each country is free to regulate the spectrum in its own way, as long as it adheres to the WARC agreement. In the U.S., two different agencies are responsible for this regulation. All the spectrum allocated for use by the federal government is regulated by the National Telecommunications and Information Agency (NTIA). The NTIA is responsible for the assignment of frequencies to agencies such as the Armed Services, FBI, CIA, *etc.* One of the most interesting aspects of how the Federal Government has regulated its own users is that communication between agencies is as difficult (if not more so) as in the private sector.

Even military channels used by the various branches of the Armed Services are different enough that they need special command-and-control radios to communicate between services. A major logistical problem during a military engagement is frequency use and coordination. Incidents where U.S. planes bombed U.S. troops occur because while both groups have radios, they cannot talk directly to each other and identify themselves; by the time the proper coordination and identification has been made through the various command channels, the damage has been done.

The federal government has access to more than 30 percent of the available spectrum. The balance of the spectrum is regulated by the Federal Communications Commission (FCC). The FCC regulates all of the radio spectrum used by the private sector. This includes AM, FM, and TV broadcast stations, public safety radio services, business and paging, amateur radio, and even devices such as police radar systems. In short, any device that transmits a radio signal in the spectrum is regulated in some way by the FCC.

The FCC is also responsible for regulating devices that generate radio spectrum signals but are not communications devices. For example, desktop computers are not radio transmitters, but they can and do generate noise on the radio spectrum. The FCC's role in the case of PCS is to regulate the amount of "spurious emissions" that can leave the case of the computer to prevent those transmissions from interfering with other devices.

Frequency Disarray

A look at any of the frequency allocation charts will reveal that each band contains allocations for most services. This is because technology has expanded the upper limits of the spectrum. Frequencies considered useless only a few years ago are now viable communications channels. Over the years, the FCC has reviewed each new chunk of spectrum as it became available and has juggled requests for this spectrum from the pool of all available users.

This is why police radio systems operate in the 30–50-MHz band, the 150–174-MHz band, the 450–470-MHz band, and now in the 800–900-MHz band. Paging systems, business radio users, and amateur radio bands can also be found in the same regions. Had all of the spectrum been available for allocation at one time, a very different scheme of channel allocations would have resulted, with all pubic safety, for example, assigned one continuous frequency band.

During the next decade, the FCC will be wrestling with what has become known as "refarming" the radio communications spectrum. It will be an attempt to review all previous allocations in an effort to try to bring some order to the allocations, as well as to change some of the channel spacing to make better use of the spectrum already allocated. This will be a major undertaking and will not be accomplished easily. Each group of users has its own lobbyists in Washington DC, and each will be fighting for a larger share of the spectrum pie. They will each try to convince the FCC that any drastic changes in allocations as they exist today will have an adverse monetary impact on their particular industry.

Along with refarming efforts, there is also a move within Congress to force the federal government to release up to 200 MHz of its spectrum for re-allocation to the private sector. This action will also take time, and it will be opposed by most of the government agencies that currently use this spectrum.

PCS and Other Allocations

One of the most recent actions by the FCC was the 1993 Rule Making creating the Personal Communications Service (PCS) in the 1.8 to 2.4 GHz band. Two things are unique about this allocation. The first is that this is the largest amount of spectrum ever released at one time for direct public access. Second, this spectrum is already heavily used by point-to-point microwave systems carrying everything from telemetry to 911 emergency calls. All of these point-to-point systems must be relocated, or the PCS providers must prove that they can co-exist on the same band without causing interference. Most of the existing users are willing to move, but the FCC has indicated that PCS providers must pay the costs associated with such moves. The exact cost of a mass movement of microwave systems to higher frequencies is unknown, but it will be billions of dollars, and will be a major concern and cost factor in the roll-out of PCS systems.

Frequency Opportunities

Even with the PCS allocations and the possible release of some of the government spectrum, the FCC will have more requests for spectrum than it has spectrum to allocate. As it progresses, refarming will help but there will never be enough of this limited resource to satisfy all those who want to make use of it. One of the ways in which some relief could be gained would be to limit the amount of time the VHF TV stations could continue to operate on their channels. At present, channels 2 though 13 occupy 72 MHz of prime spectrum. As we move forward with more cable access, direct broadcast satellites, and HDTV, if these stations had to move to the UHF band, this region of the spectrum could be re-allocated for use by the general public.

The FCC has the unenviable task of trying to balance the requests for spectrum against the amount available. Now there is another issue it must address—that of raising money for the General Fund. PCS allocations are the first that will be made available to potential system providers through an auction process. The federal budget already includes the billions of dollars these auctions are expected to generate, and the FCC will continue to find ways to make the spectrum pay for helping to run the government.

This new attitude of charging for a scarce public resource is designed to provide a new form of revenue for the government and to prevent the profiteering that accompanied the cellular licensing by lottery. In the case of cellular licenses, people who had no intention of building a cellular system entered the lottery, were awarded a license, and almost instantly sold it for millions of dollars. The FCC and the government would rather have the revenue come directly to the government.

There is no easy solution to the spectrum shortage situation. Technology advancements will continue to help utilize spectrum more effectively, but the demand for sending and receiving more information at faster speeds will negate such advances. Spectrum will always be in short supply, significantly limiting our ability to provide a go-anywhere, do-anything, access-anything wireless world.

CHANNELS AND BANDS

A channel is a specific frequency assignment within a band. The radio station KGO in San Francisco is assigned a frequency of 810 KHz (their channel) within the AM radio "band." Therefore, a "band" is a group of channels or frequencies used by the same service or type of service within the industry.[1]

Television channels 2 through 6 occupy the 54 to 88 MHz range (channel 1 is not used in the United States, instead the frequency allocated for channel 1 is occupied by the amateur radio 6 Meter- (50–54-MHz) band). FM broadcast stations utilize 88 to 108 MHz, and television channels 7 through 13 are assigned frequencies starting at 174 MHz and ending at 216 MHz. (As a point of reference, each television channel occupies 6 MHz of spectrum. Because of interference problems, no city has two channels assigned side by side. A city with a station on channel 3 will not have a station authorized on either channel 2 or 4; however, since there is a "gap" in frequencies between channels 6 and 7, it is possible for a city to have both a channel 6 and 7 television station in operation.)

RADIO SPECTRUM

The Electromagnetic Spectrum, from Very Low Frequency (VLF) to Gamma-Ray is divided as in Figure B-1. The Radio Spectrum is futher divided as shown in Figure B-2. The portion of the Radio Spectrum used for Wireless Data is shown in Figure B-3.

---

[1]A meter is 39 inches and it is the measure of a wavelength at a given frequency. Therefore, the 80-meter band has a wavelength of 80 meters or 260 feet. Generally, the reference to bands in terms of meters is left over from yesteryear, but it is still used by amateur radio operators. In the world of two-way radio, references to bands are now prefaced with the frequency range with the terms High Frequency band (HF), Very High Frequency band (VHF), and Ultra High Frequency band (UHF). Above the UHF band, the designator is the frequency range: the 800-MHz band, the 2-GHz band.

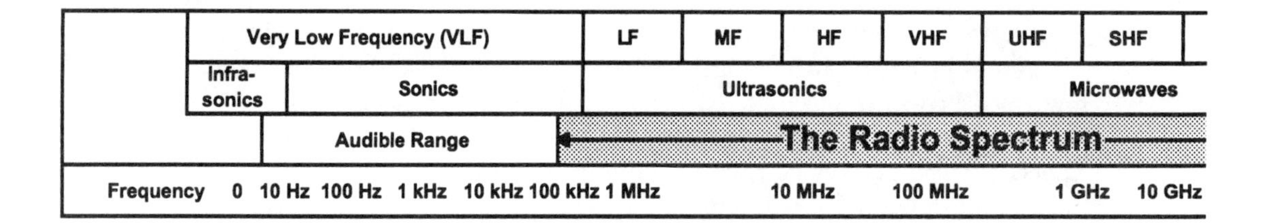

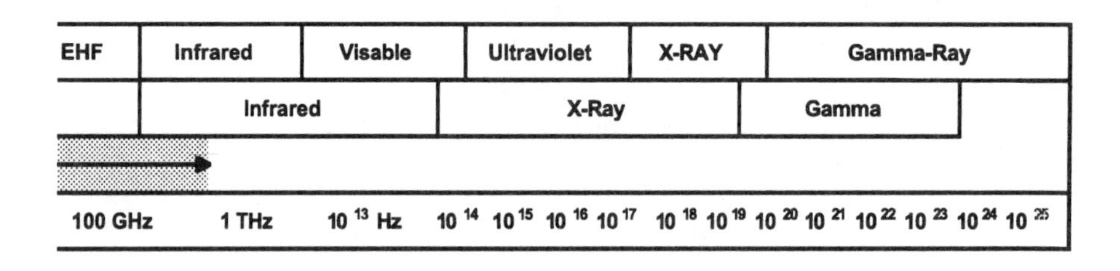

Figure B-1. The Electromagnetic Spectrum

Very Low Frequency (VLF)		LF		MF	HF	VHF						UHF	
Maritime Services	Nav	AM Radio	Int. Comm.		C B	BR	TV	FM	Air	BR	TV.	Gov/Ham/PS/BR	
Frequency 10 kHz 100 kHz 1 MHz 10 MHz 30 Mhz 100 Mhz 150 Mhz 200 Mhz 400 MHz 500 MHz													

UHF				SHF	EHF	Infrared
	UHF-TV	Cell/SMR	G	PCS	Microwave	Infrared
		800 MHz 1 GHz 2.4			2.5 GHz 10 GHz 100 GHz 300 GHz	

Figure B-2. The Radio Spectrum

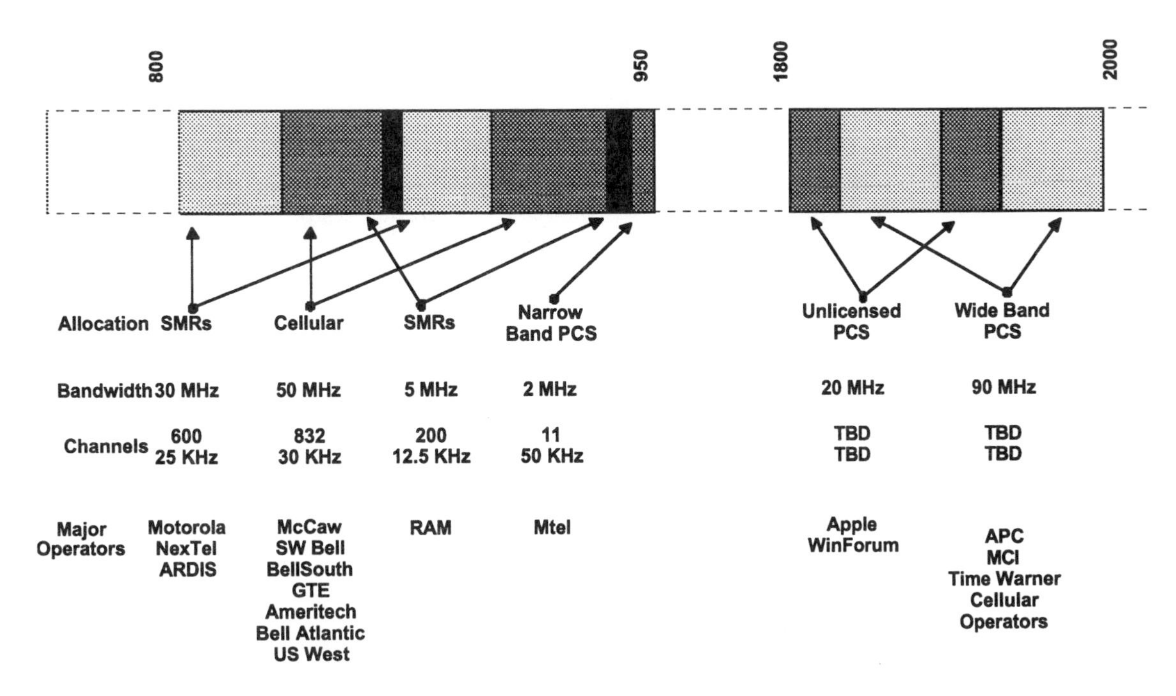

Figure B-3. Wireless Data Spectrum

MODULATION

Present day transmission technologies will not, for the most part, handle the data speeds required by the new wireless information devices.

A radio link can be considered as though it were a wired link for purposes of data transmission speeds—the smaller the wire, the lower the data speed (smaller wire equalling narrower frequency bandwidth). There are several ways to increase data speeds. One is to compress the data. Another is to packetize it and transmit it in bursts. Yet another is to increase the bandwidth.

Amplitude Modulation (AM)

To put the issue of bandwidth into better perspective, consider standard AM, FM, and TV transmissions. The term "AM band" describes the type of audio being transmitted. In this case, we are discussing "Amplitude Modulation." Our standard AM broadcast stations are located in the band from 530 to 1600 KHz. Stations can be assigned at intervals of 10 KHz within this band. However, the FCC takes great pains not to allocate adjacent channels to stations in the same broadcast area. Therefore, a station might be found on 530, another on 610, and another on 810 KHz in one area, but one would not be found on 530 and another on 540 KHz.

In an AM broadcast transmitter, the frequency stays the same (i.e., 610 KHz), and the audio signal is superimposed on it (Figure B-4). This audio signal is modulated up to the 85–95 percent range. The frequency of the station is not permitted to vary more than 20 Hz above or below the assigned frequency at any time, and the station must maintain its transmission so that no side-band distortion is interjected into the signal.

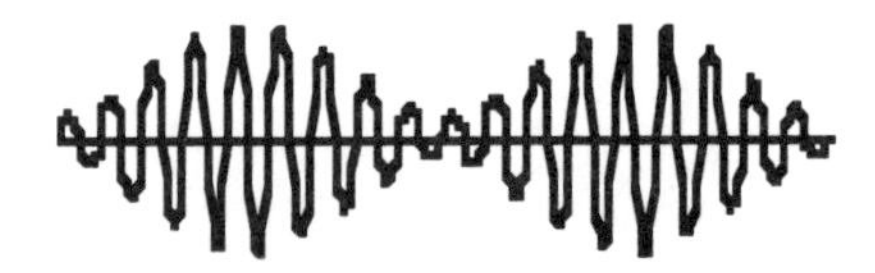

Figure B-4. *Amplitude Modulation*

Frequency Modulation (FM)

The FM broadcast band (88–108 MHz) consists of 100 allocated channels. Each channel is assigned 200 KHz of bandwidth, and in a given area, the channels are separated by one, two, or three vacant channels, depending on the physical distance between the stations' transmitters. FM stands for "Frequency Modulation" and it means that instead of the audio being superimposed on top of the radio channel, the frequency of the channel actually changes with the changes in modulation (audio) applied to it (Figure B-5). A station operating at 107.5 MHz, for example, is centered on that frequency. In reality, the station's frequency is shifting from a low of 107.425 to a high of 107.575 MHz (+/- 75 KHz). When no audio is present, the radio carrier is centered at 107.5, but when audio is applied to the carrier, the radio frequency changes both plus and minus. This change occurs first in the positive direction and then equally in the negative direction. Without getting too involved in this technology, note that FM broadcast stations also transmit what are referred to as sub-carriers. These sub-carriers are used to transmit a second channel (for stereo), or services such as Muzak, or even data transmissions. (Epson ran a stock quote system on FM sub-carriers for a year or so before canceling the project.)

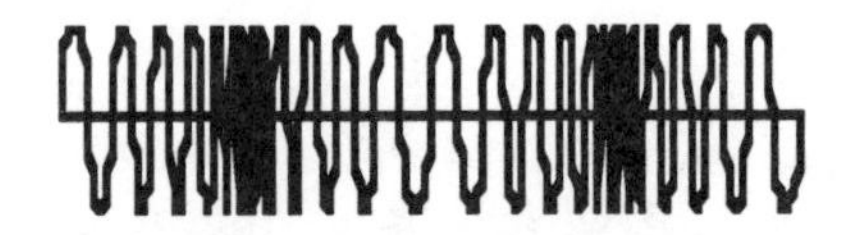

Figure B-5. *Frequency Modulation*

Since FM broadcast stations use a form of frequency modulation occupying a total of 150 KHz (+/- 75 KHz), the signal is referred to as Wide-Band FM (WBFM). (Each channel is 200 KHz wide and +75 to -75 equals 150 KHz. The remaining 50 KHz is the buffer zone.) Two-way radio and paging transmitters are limited to +/- 5 KHz of bandwidth. This does not affect the distance or range of the transmitter, but it does limit the audio range that can be used. Two-way radio and paging systems do not need "broadcast" quality audio and the +/- 5 KHz bandwidth was found, over time, to be a good compromise. It permits more channels or frequencies per megahertz of radio spectrum, yet it permits normal voice communications to be received and understood. In the space reserved for one FM broadcast station, it is possible to fit about fourteen separate two-way radio or paging channels.

TV Bandwidth

The hog of the spectrum is the public television network (TV). Each television transmitter uses 6 MHz of spectrum. Thus, the same amount of spectrum used by one television channel would provide space for 30 new FM broadcast stations, or 400 two-way radio or paging channels. TV channels 2–6 occupy the frequency range of 54 to 88 MHz, channels 7 through 13 from 174 to 216 MHz, and channels 14 through 69 (UHF) from 470 to 806 MHz. This is a total of 412 MHz or 2,746 additional two-way radio channels.

The TV transmitter is a complex device. The video and audio signals that arrive at our television sets are transmitted in a portion of spectrum that is 6 MHz wide. Some of this space is occupied by the video signal, some by chrominance subcarrier frequency (color), and some by audio signals. In addition, there are several other types of information that can be sent along with the video. Some stations simulcast digital signals that are converted as printed text for the hearing impaired. Stereo audio channels can be provided, and some side-bands can be used to transmit other forms of data.

Two-way Radio

Again, existing two-way radio systems use frequency modulation that is limited to +/- 5 KHz. In converting this to data capabilities, it is probably best to consider that a single two-way radio channel can handle data at about the same speed as a standard telephone circuit. Thus, at present, the limitations are about the same. A standard telephone modem can transmit and receive data at a speed of 9,600 bps. With various compression techniques, this speed can be increased to 14,400 or even 19,800.

This translates to textual data that appears on a computer screen faster than it can be read, but the rate of transmission is not fast enough to enable graphics and large amounts of data to be passed without consuming a large amount of time. To put this in perspective, current Ethernet and Token Ring data rates are a minimum of 1 Kbs (kilobits per second) and up to 14 Kbs or higher. Those who think a wired network is slow, should imagine how slow it would be if data rates were cut by an order of 5 or 10 times.

Another way to visualize the bandwidth versus data rate issue is to picture a garden hose with an inside diameter of 1 inch. Turning on the faucet will cause water to flow through the hose and come out the other end. The rate of this flow can be increased by increasing the amount of pressure behind the water (opening the faucet wider). At some point, however, no increase in water pressure will be delivered to the end of the hose. The hose does not expand to permit more water to flow through it, and once a certain optimum pressure level is reached, increasing the pressure input will not force more water through the hose.

In order to be able to send and receive data at speeds that make sense, RF bandwidths have to be increased, compression technologies have to be improved, or technology has to advance to permit more data to be passed over less bandwidth. The reality of the situation is that all of the above are being tackled by the industry. There will be some advances in the amount of data that can be transmitted over a given bandwidth, but the demands placed on the systems by expectant users will not be met.

Technologists involved in wireless communications are attempting to do exactly what wireline technologists are doing—send more information down a pipe that remains the same size as that previously used. This works to a certain point, and then the laws of physics take over, just as they do in the case of the garden hose. Where is the limit? LAN speeds over twisted-pair wires have increased from a few hundred bits per second to 14,000 bps. In wireless communications at the moment, however, the physical limit to the speed due to the limited bandwidth has been reached.

Just as wireline companies are looking toward optic fiber cable because it has a much greater capacity, data engineers are looking at the radio spectrum and asking for allocations with more bandwidth. Another illustration of the current situation is the difference between the single pair of telephone wires and the TV cable. Over the single pair of telephone wires, only a single telephone conversation can, at present, be in progress. The TV cable has the capacity for 120 or more channels of TV programming. A voice conversation on a telephone is like a two-way radio conversation—the audio quality is good but not great. Yet, a cable only a few feet away carries 120 channels of TV representing about 720 MHz of radio spectrum. The difference is in the cable used. The telephone company could change the phone system today and put several more calls on the same pair of wires, but each added conversation would limit the audio characteristics of the other calls in progress. Soon users would get to a point of not being able to understand the person at the other end.

The TV cable, on the other hand, is really an RF antenna system. Since it is not running through the air, all of the frequencies allocated by the FCC to other services can be used inside the cable. The 120 channels received are "transmitted" on the cable at RF frequencies and then "tuned" in by the cable box or cable-ready television set.

MORE WATER INTO THE HOSE

A technical search is under way for better ways to cram more water into the hose, making sure it all comes out the other end. This may be an oversimplified way of looking at the problem, but what is going on in the world of wireless is that proposals in front of the FCC for special radio spectrum allocations for the Personal Communications Services (PCS) are attempts at providing larger hoses. New transmission technologies are addressing the amount of water that can be passed through the larger hoses.

Many different types of hose will be necessary, however: one or more for sound (e.g., voice), one or more for data, and several for video systems. The FCC is struggling with the problems associated with reallocating radio channels and making them available for the new PCS. Technologists, meanwhile, are trying to find new and better ways to move more data faster.

Digital Systems

There is no doubt that converting typical analog systems to digital will increase both the data speeds and the amount of data that can be transmitted. Simply converting the existing analog cellular phone system to digital in the United States will increase the number of calls that a system can handle by as much as 40 percent. In a purely digital world, the movement of voice, data, and video becomes much simpler. Data—that is, bits that are on or off—can be moved at greater speeds, mixed up, encrypted, verified, and then reassembled at the other end in such a way that the receiving station never knows if bits of data were dropped and had to be re-created. Voice and information (print characters, graphs, etc.) can be multiplexed on the same cable or radio channel, sent to their destination, and then converted back to voice and information bits without anyone knowing they traveled together on the same radio carrier.

SPREAD SPECTRUM

One new technology being employed in wireless communications is called "spread spectrum." This has become especially popular in the 902–928-MHz radio band. This band is available for low-powered transmitters on an unlicensed and, therefore, uncoordinated basis. It can be used by amateur radio operators, government, garage door openers, experimenters, cordless telephones, wireless LANs, and more. Since the systems are all low-power, there should not be much chance for interference. Using spread spectrum technologies, there is even less of a chance for interference. However, since this band is not regulated and does not require a user to have a radio license, there is no guarantee that a system that works in this band today will continue to function properly as more and more transmitters are put into use in the future.

Several attempts have been made by the communications industry to explain spread spectrum—each leaves the reader more confused as he or she wades through the techno-jumbo. No wonder the computer and communications industries do not understand each other. Using the garden hose example, suppose instead of a hose that only had a single end to it, a garden sprinkler hose—the type that is full of holes, and is designed to wet a large garden area without having to spray large amounts of water—is used in the example. Now suppose the holes along the length of the hose could open and close in a random manner. Each time water is let into the hose, the holes would open one at a time, very quickly, letting water out a little bit at a time. That is one form of spread spectrum. The other half of the equation is that the receiving end (the flowers) would know exactly which hole was going to open when and would be waiting there to catch the water as it came out. The knowledge of "which hole is going to open when" is the key to the success of frequency-hopping spread spectrum. The idea is that instead of transmitting at full power on a single channel, users transmit for very short periods of time on many different channels. This type of technology has been used by the military and others for many years. It is not a new methodology; it has only been applied differently in this case.

The other form of spread spectrum uses a single transmitter with a very wide bandwidth. As the signal is modulated, it is sent out over a portion of the transmitter frequency and received at the other end. The equipment does not need to know exactly what frequency is in use since it processes the entire channel, filtering out the noise and recovering the data.

SYSTEMS INTEGRATION

Any computer is made up of sub-components, each requiring integration into the system to provide a comprehensive unit. Such sub-components include a power supply, main processor, memory, display, and input/output assemblies. Each may be a single chip or a combination of chips and/or other fairly generic electronic components. Each is designed and constructed on printed circuit boards (PC boards). Components are mounted on the boards and conductive lines that replace wires connecting one component or sub-segment to another are etched on the board.

As computers become more sophisticated, printed circuit boards become more difficult to lay out. Each "trace," or etched conductive line, must connect only the elements it is intended to connect. It cannot touch or cross other components or traces it is not designed to connect. The difficulty inherent in this process cannot be overlooked. Consider a series of highways and smaller roads. It is important to design these roads so that they intersect only at the proper places, yet some roads need to cross highways without making a connection. In the world of road construction, this is handled by building overpasses and underpasses to rout one road over or under another. In complex interchanges, these become a series of ramps and bridges, each designed to direct a specific road to a specific point while keeping it separated from the others.

Over the years, electronic engineers have developed their own sets of bridges and interchanges to handle the electronic pulses and voltages that must travel over the circuit boards. Instead of bridges, this is accomplished on a PC board by adding layers to the board, one of top of another. Each layer can be laid out in such a way as to provide the best and most direct path for component connections. Layers can be built on top of each other, using the printed circuit board material to keep the traces on one layer from coming in contact with those on another layer.

Today, multiple-layer printed circuit boards are the rule rather than the exception. Every computer makes use of this technology. The more complex or smaller a computer is, the greater the need is for layered construction. Once the computer has been designed and built using these techniques, it must pass a set of FCC standards for radiation and spurious emissions. Many of the components used inside a computer can "radiate" signals that may interfere not only with other components on the same board, but, under certain conditions, they might also create interference for devices located near them. Therefore, the FCC has set up a series of guidelines requiring computer vendors to design their systems to minimize stray radio frequency emissions that could interfere with other equipment.

The computer vendors of the world have responded to these requirements well. Most, if not all, computers presently on the market meet or exceed these minimum interference requirements. Thus, computers that do not radiate unwanted radio frequency signals (RF) have been developed. Wireless devices are designed and built to even more rigorous standards. So, what's the big deal with putting them together? The answer to this question is that in addition to controlling radiated radio frequency signals, computing devices must also be protected from being interfered with by other devices that may radiate radio frequency signals. And a two-way wireless device is designed to radiate these signals!

Some of the same precautions used to prevent a computer from radiating RF signals also help to protect them from interference caused by other devices, but the level of protection is not enough to overcome the problem.

To appreciate some of the complications associated with the marriage of these two technologies, it is important to understand how transmitters and receivers work. More detail is provided below, but, in short, a communications device must be connected to an antenna to function properly. An antenna in its most basic form is a simple wire that has a length relationship to the signal it is designed to receive.

This can be demonstrated using a car radio and antenna system. When an FM broadcast station is tuned in with the antenna fully extended, the radio signal will be received at maximum strength. If, however, the antenna were to be pushed down or collapsed, the radio will start losing the signal, and the result will be weaker reception.

Remember that the traces on a printed circuit board are used to replace wires that must connect the various components. Since these traces are really wires, if their length is just right, they will act as an effective antenna for a transmitted signal, feeding that signal back into the components to which the traces are connected. If the signal received by the PC board trace is strong enough, it can either damage the circuit or impede the signals that are supposed to be traveling on the trace, thus interfering with the proper operation of the circuit.

Receivers, while designed to receive a signal and convert it to information that can be understood (voice) or used by a computer (data), also emit radiation that can interfere with a computer. Just as a wireless transmitter can interfere with a computer, so, too, can a paging or messaging receiver.

Perhaps the earliest instances of these potential problems manifested themselves in automobiles. As car engines, braking systems, and even climate controls became more sophisticated, auto manufacturers began using computers as miniature "brains" to control various aspects of a car's operation. In the mid-1970s, the Cadillac was one of the first autos to incorporate anti-lock brakes. These brakes are designed so that when the driver applies the brake hard, the circuit will prevent the brakes from locking up, causing them to pulse on and off instead.

The Ohio Highway Patrol cars were equipped with high-powered mobile units (100 watts) operating on their FCC-assigned channel. The first time the issue of radio interference affecting automobile computers arose was when an Ohio Highway Patrol officer started transmitting while driving at 60+ miles an hour on a freeway. The Cadillac driving next to him suddenly started bucking and jerking erratically. The officer pulled the car over and started to admonish the driver for driving in an unsafe manner. The driver tried to explain to the officer that he had done nothing at all; suddenly the car had started acting as if the brakes were being rapidly turned on and off.

This first instance was followed by many more until someone noticed a correlation of the transmissions with the events. It was particularly difficult to discover the true cause of the problem since it only happened in Ohio, and only on major highways. There were no reports of the Cadillacs misbehaving anywhere else in the nation. Once the problem was identified, Motorola and other communications suppliers began working with automobile manufacturers to make sure that as more and more computers were included in cars that they were properly shielded and that RF problems were minimized. However, even today it is not unusual to hear stories about cellular phones causing interference with a car's computer ignition or other computerized system.

To add to the complexity of these issues, even when a specific computer has been optimized for a specific radio or group of radio channels, putting a different radio on a different frequency in the computer may mean starting the process all over again. Since the wireless data networks operate on diverse radio frequencies, it is not enough to optimize a system for, say, RAM and expect it to work properly on NWN or even ARDIS. In many cases, there might be interference. If the user is working with a packet system such as RAM or ARDIS, he or she might not be aware of the interference—the system might seem to be very slow in sending and receiving data.

The two modem/transmitter/receiver combinations available for wireless data communications today are external units that connect to a portable computer through a serial connection. These units are made by Ericsson (private labeled by Intel) and by Motorola. Both units are about the size of a small handheld computer.

These units transmit on either the RAM or ARDIS wireless network at output powers of up to 2 watts.

There have been no reports of any cases of radio interference from or to a handheld computer using either of these units. However, it is still possible that some types of interference might occur if a given computer is not properly shielded or if other external devices are added. (Even a standard telephone cable used for a wired modem can function quite effectively as an antenna, causing interference to a unit.) If the transmitter is causing interference to a computer's normal operation, it should be evident. However, some forms of interference might show up as slow network response or delays in receiving messages. Since these devices are using packet-based rather than session-based networks, the end user might not be aware that there is an interference problem—only that messages seem to take longer than normal to be sent or received.

The real challenge will be with the tighter integration of the radio transmitter/receiver unit and the computer. PCMCIA-type wireless modem units being developed by Motorola and others will be inserted into the PCMCIA slot where they may, in fact, cause additional types of interference.

Computers that will work with such internal wireless modems are going to have to be designed to minimize the potential for radio interference. Even the placement of the PCMCIA slot will be of utmost importance. If a system's CPU (Central Processing Unit) or memory chips are placed directly above or below the slot, the potential for interference with the computer is greatly increased. If the wireless modem's power leads are not properly shielded, and if the computer's power supply is used to provide power to the wireless modem, RF can travel down the power leads and cause a variety of problems.

The key to success for the integration of radios and computers lies in each set of design engineers working on their own piece of the puzzle with an awareness of interference potential. They must make sure to minimize the chance that stray RF will get into the workings of a computer or that RF signals being broadcast by a CPU will impact the wireless modem's ability to receive and decode a signal intended for it.

The next generation of equipment—devices that will be built from the ground up to support both computing and wireless data transfer—will require that computer and communications engineers work together closely to ensure that the systems are properly integrated and that the potential for interference is minimized.

Moreover, since there will be many different types of systems—operating on many different radio channels—the integration of computing and communications will be an on-going process. As each new processor with its higher clock rate and increased performance enters the market, and as each new breakthrough in communications technology reaches the market, these devices will require more and better design considerations when being integrated on a single board and sharing a single power supply.

Many communications engineers are prepared to tackle these issues, having already mastered some of them through the design and construction of voice cellular phones which feature embedded CPUs, and even pagers where the "brain" of the unit is a CPU working in conjunction with the receiver.

Radio interference is well understood, but is often difficult to isolate as the cause of a problem. This is especially true when the transmitter is an integral part of the device. There are a number of different frequencies in use within a computer. A system with a CPU running at 50 MHz has a reference crystal in it that is operating at a lower frequency. This frequency is then multiplied up to the 50 MHz required for system operation. If there are several stages of multiplication, as there are in radio receivers and transmitters today, a system may generate radio interference at any or all of these frequencies plus the sum and/or difference of any two or more. This phenomenon is referred to as the "harmonics" of a system. For example, it is possible for a system's clock running at 25 MHz to be doubled to produce a 50-MHz signal. But the circuits used to multiply the 25-MHz clock to 50 MHz might also generate RF noise on 12.5 MHz (1/2 of the original), 50 MHz (the final desired output), 75 and 100 MHz (sums), and others. It is then possible for two or more of these frequencies to "mix" or combine within the unit and introduce additional radio noise back into the system at other multiples of radio frequencies.

Radios, too, generate multiple frequencies in addition to the specific frequencies they are designed to generate or receive. As an example, a typical receiver designed to operate in the 450–470-MHz band starts out with a reference oscillator (crystal) on a frequency of 14.4 MHz. This is then divided into two different frequencies—4.166 KHz to provide a reference frequency, and another that is used to create the actual operating frequency.

At the front of the receiver, the portion of the radio frequency that needs to be received is amplified and narrowed to the desired frequency as determined by the synthesizer. It is then mixed with the synthesizer frequency that is set to 10.7 MHz below the actual received channel, filtered, and amplified again. Finally, the audio is "peeled" off the signal and processed.

For those who do not intend to design radio receivers and transmitters, it is not important how all this happens, only that during the process there are many different frequencies generated and used. A receiver can be interfered with not only on the channel it is listening to, but also by a signal on many other frequencies as well.

A radio transmitter also generates a number of different frequencies on its way to providing a specific output frequency. The simplest example of this is, again, a 450-MHz transmitter that uses a reference crystal of 12 MHz and then, through a series of frequency doublers and triplers, arrives at the operating frequency. Although limited by FCC rules and regulations, there are still components of the 12-MHz reference and all the combinations of the doublers and triplers "leaking" out of the radio that could cause interference to a receiver or computer in close proximity.

The point of this discussion is that the marriage of computers and communications devices is not a simple one, nor is it an exact science. While computer models can be run showing the relationships between various radio frequencies and cable lengths, until a system is actually built there is no guarantee that it will not be susceptible to radio interference of some type. And even after the system is functional, a change in conditions such antenna length or placement—or any number of outside influences—can change the parameters and create a problem where no problem existed before.

An important point here is that radio waves are a lot like colors in that they can be combined or subtracted to create other frequencies (colors). As users become more mobile, equipment designers must be aware of all these possibilities and design equipment in such a way as to guard against such problems.

Outside influences will indeed play a role in these design criteria. For example, what if, while using a RAM wireless modem to send and receive messages, a user decides to check in with his or her office over a handheld cellular phone? In turning on the cellular phone, the user has introduced additional radio frequency signals into the environment. These signals are close in frequency to those used by the modem. It seems that it should be easy enough to test equipment to make sure that it will work in this type of environment, but any one of the thirty channels licensed by RAM and any one of the channels on a cellular phone (666 or 832 depending on the phone) can be used, and that these channels could change as the user moves around, the reader will begin to appreciate the enormity of the task.

It is not impossible—it will work. But it is vitally important that the basics of what is being attempted are understood and there are many variables. Consider for a moment the person at a company help desk who gets a call from a user. The user describes a problem he or she is having with one of the company's computers but fails to mention that there is a PCMCIA wireless modem inserted in the PCMCIA slot. There is no way in the world that the help desk attendant is going to be able to assist the user unless there is at least a general level of understanding that a communications device can interject a variety of conditions into an otherwise perfectly working computer.

If a vision espoused by some within the industry is correct, many users will soon be carrying a handheld computing device and several PCMCIA cards. Within the walls of their company, users will insert a wireless LAN card and connect to their company's LAN. When they leave the building, another wireless modem will be inserted to be able to connect to the wide-area service provider.

In another instance, users might carry a device containing two separate transmitters and receivers. When they decide to buy one of these devices and go to the shelf marked "wireless modems" and choose their service provider and type of wireless network, they need to have some confidence that each of these devices has been tested with any of the computers they might decide to use and that the computers will work well together.

Awareness is the key to our wireless future. It is not necessary to understand everything there is to know about wireless receivers and transmitters; users only need to be aware that combining a computer with a communications device will add to the complexity of the overall system. If a problem does arise, users must be able to have enough of an understanding of the relationships so they can isolate and correct the problem easily.

SIMULCAST

When most people hear the word "simulcast," they think of one of two things. We have all heard the term used in conjunction with concerts that are televised with the sound portion being sent over an FM broadcast station to be received in stereo. Or, two radio stations—one AM and one FM—use simulcast to reach larger audiences. Both uses of the term are correct, but neither is anything like simulcasting technology used in one-way and two-way radio systems.

Both of these forms of simulcasting make use of different frequencies or channels. The audio is sent to two different transmitters, each on a different channel, and the listener decides which to tune to. Simulcasting in two-way radio is very different, and vastly more complex.

In two-way radio, "simulcast" means that two transmitters, each tuned to the same radio channel, transmit the same information at the same time (Figure B-6). This sounds simple enough, but it is made more complex because of the nature of radio signals.

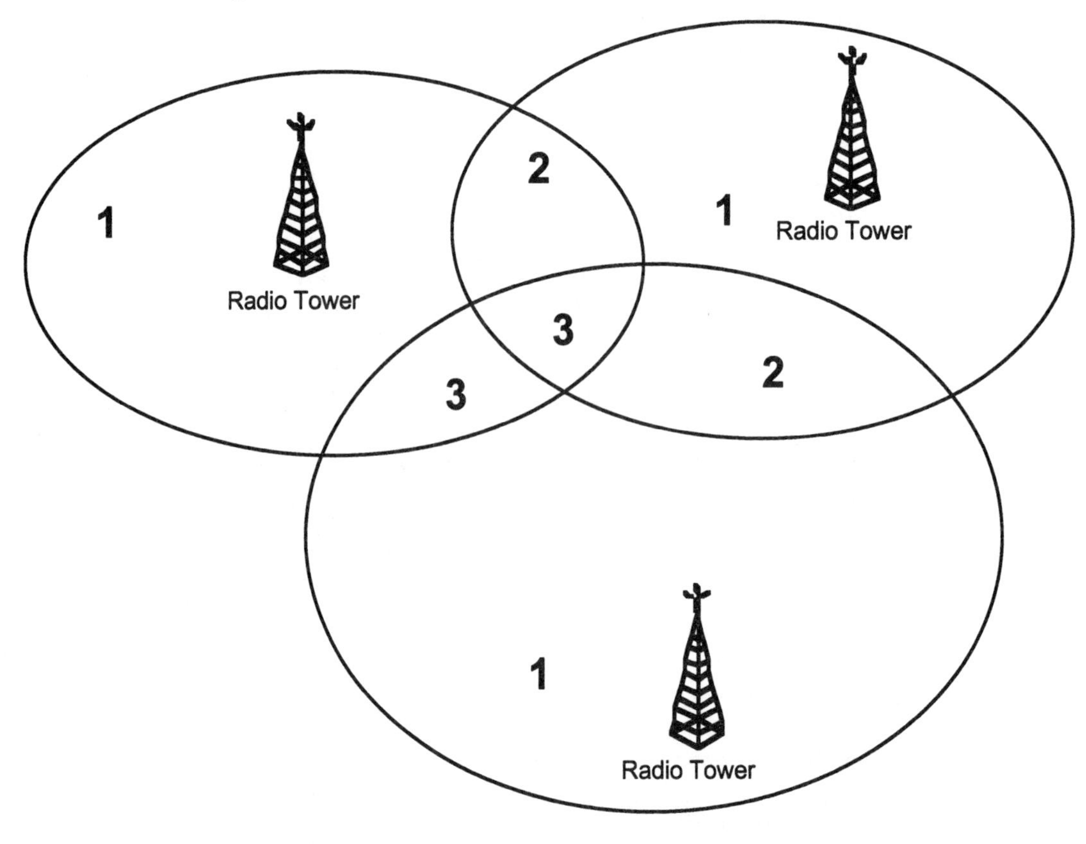

Figure B-6. *Simulcast System for Two-Way Radio (Wireless Systems)*

To understand how complex this technology can be, try to listen at night to an AM radio channel that carries more than one station.

During the day—because of the characteristics of radio wave propagation—only one station will normally be heard. At night, however, when propagation characteristics change, it is possible to hear two stations on the same channel—both the local station and one located in another city. Since the coverage of the two stations overlaps at night, a mixture of the two stations can be heard. Most of the time, listeners can not understand what is being said on either station.

Moving this example to two-way radio systems, the difference is that both stations would be transmitting the same information. A user might wonder, "If that's the case, what's the problem?" The answer is that there are many problems, only one of which is synchronizing the audio.

Radio waves from two different transmitters can arrive at a location at different times. Radio waves travel through air from a starting point (the transmitter) to the point where the receiver "hears" the information. If a user were exactly the same distance from two transmitters, he or she might hear the composite of the two signals and understand what is being said. However, if the transmitted audio is out of phase between the two, they would have a tendency to cancel each other out and you would hear *nothing*.

Or, consider connecting two sets of speakers to the same radio or amplifier. If they are out of phase, theoretically a spot could be found that is equidistant from each set where no sound would be heard. In practice, the user probably would not be able to find this point, but he or she would be able to find multiple places in a room where the sound was distorted and not pleasing to the ear.

It is the same with two-way radio, and engineers have spent many years working with techniques that will resolve this condition. If two transmitters are used to cover an area, the only place there might be a problem is within the overlapping area. In some locations, the signal from only one of the transmitters will be received. In others, one transmitter will be strong enough to block the signal of the other, so only one will be heard. But where both signals are being received with equal strength, difficulties are experienced.

If transmitters are added, two more parts of the equation appear. First, there will be more areas where transmitters will provide overlapping coverage. Second, there will now be places where three, four, or more transmitters will be heard.

To overcome such problems, many local paging services have decided not to try to synchronize their transmitters. Rather, they transmit the data stream multiple times. The data is sent to the first transmitter, then to the second, then to the third, and so on. This works well, but it is not a very effective use of channels. A three-second burst of data will tie up a channel much longer if it has to be retransmitted by multiple transmitters.

It is not necessary to delve into the engineering considerations required to make a simulcast system work; it is only necessary to understand that making it work is not an easy matter, and it adds considerable cost to the overall system. Radio transmitters can "drift" slightly up or down in frequency, audio signals can be distorted by many factors, and radio waves can bounce off objects, changing their polarity.

The point is that simulcasting is one of the most difficult of all radio technologies to get "right," but it also has some decided inherent advantages. Installed properly, simulcast systems improve and expand the radio coverage area for a given channel. If the audio is matched and balanced, a signal in an area covered by more than one transmitter is enhanced. That is, the combined signal strength is greater than that provided by a single transmitter.

In the case of the SkyPage and NWN systems, simulcast technology becomes very important. It is this additive effect that provides the superior coverage offered by SkyTel within buildings and other hard-to-cover places. It is this technology that permits seamless movement of a mobile pager with little, if any, chance of it missing a message. Increased coverage will also be provided by NWN with the use of simulcasting.

The systems that provide the best coverage will be the systems that will attract the most users. The average user will not be happy having to feel out locations where he or she can send and receive data. If NWN is as successful with its implementation of the simulcast technology as SkyTel has been, it will have yet another advantage over other service providers.

DATA TRANSMISSION METHODS

The most difficult aspect of wireless communications for the average person to understand is the difference between "session-based" and "sessionless" communications. It is this difference, and people's willingness to accept sessionless systems, that will help move the wireless industry ahead.

SESSION-BASED SYSTEMS

The reader who is familiar with accessing information by connecting to a modem and dialing out over a wired phone system to connect to a remote computer or system is familiar with session-based communications. In this example, the modem dials out over a wired telephone network and the network connects the caller directly to the device or system at the other end. Until the connection is broken, the caller is directly connected. When typing, each character is sent over the phone line to the other end and echoed back to the machine. If the user is typing a command line, it is being assembled at the remote end of the system. When the "return" or "enter" key is pressed, the command is executed.

During a direct-connect session, any noise on the phone line may be manifested as incorrect or extraneous characters. When those characters appear on the screen, the user can never be sure whether they were generated on the return trip (meaning that the remote machine did not "see" them) or if they are going to interfere with the instructions to the remote system.

During a session connection, the users become captive to the remote system. They send a command and then must wait for a response. They must stay connected to interact with the remote system; if the connection is interrupted due to excessive line noise, or a dropped connection, the user must start the session over again.

Most users of remote computing are accustomed to this type of interactive remote access and tend to think that it is the best method to connect to a remote system. When the word "packet" is introduced into the conversation, most modem-proficient users do not understand the concept, nor do they understand that it is a much more effective method.

Packet Systems

The best way to describe the differences between session and sessionless communications is in two steps. The first is to discuss normal Local Area Network (LAN) connections, and the second is to apply the concept to the world of wireless.

In a LAN, the systems are interconnected with a wire of some type. It might be telephone cable or a piece of coax, but all the computers on the network are connected by cables. The transmission scheme used on *all* the networks is based on sending and receiving packets of data. This is because if the network had to provide session connections for users, only a single user at a time could use the network. By using packets of information, any number of users can send and receive data across the network without any noticeable degradation or slowdown. Though it works differently from session connect systems, the results are the same.

The user still enters a command on his or her screen. However, instead of each keystroke being sent to the remote computer, the entire command is first typed on the local screen. When it is complete, the command is packetized and sent to the remote device where it is acted upon. Depending upon the system and protocols used, a packet of data can range in size from a few bits to several thousand. The actual length of the packets is not as important as the information they contain.

Each message is divided into a number of packets. Each packet contains not only a portion of the message or command, but the address of where it is to be sent, the address of the originating node, the number of packets comprising the total message, and error correction codes so that the receiving unit can verify that it received exactly what was sent. Once all the packets have been received, the routing information is stripped out of each packet, the packets making up the message are reassembled, and commands are executed by the remote machine.

The reason for using this type of transmission is that messages from a number of different nodes can be traveling along the network—more than one user at a time can access the network. If a user types a message that is broken down into ten packets, and another user sends one at the same time that is also ten packets long, the packets can be interleaved on the network—that is, the first packet from user A is sent, and right behind that, the first packet from user B enters the network, followed by the second packet from user A, etc. Because each packet is numbered and contains all the addressing information, the packets are reassembled at the remote end in the proper sequence. The packets from the two users do not get mixed up.

Packets on Wireless

The same principle applies to packets over wireless networks. A single-channel wireless network can be compared to a wired LAN. Packets entering the network from various users, intended for different destinations, can co-exist on the channel without causing interference to each other. The other advantage of using packets is that the transmitter on the handheld system does not have to stay on for long periods of time. It sends short bursts of packets and shuts down until the next packet is ready.

Cellular voice and analog data connections, on the other hand, generally require the device's transmitter to stay on during the entire conversation, even during any pauses. This tends to deplete the battery more quickly.

What the end user experiences is both similar to and different from a direct-connect session. First, on the screen, the user sees the command or message. After it is sent, the answer or the result of the command is observed. What the user does not see are any characters generated due to line or system noise, or any retries and resultant garbage that accompanies them.

The other point that is confusing to many experienced with direct connect communications is that they are generally infatuated with the speed of the modem; it is important that they have the fastest modem available. In reality, a 2,400-baud modem does not provide full data transfer at 2,400 baud, nor does a 9,600-baud modem provide a throughput of 9,600 baud. Even with wireline systems, the actual speed of the data is somewhat less than the rating of the modem.

In the packetized world, the throughput is slower—given the same speed modem—because of the overhead associated with each packet. But raw speed becomes immaterial in a packet-based system. All the user knows, or can measure, is the time delay between sending a command and the result being displayed on the screen. Speed, as it is used to measure modem performance, is a non-issue in the case of packet.

Packet communication is a more viable and reliable form of wireless communications than direct session connect is, and packet communication shields the user from any transmission problems encountered on a network. However, packet is not the optimal method for sending large graphical files or large amounts of data. Different types of networks should be considered for different purposes, and users have choices to make. For example, users who have elected to have a CDPD-only modem to make use of packet technology over cellular systems will experience difficulties when trying to send and receive fax documents in their native graphical form or other large files. Users who need to be able to send and receive faxes and other large documents should opt for a modem capable of both packet and analog data.

CONCLUSION

Many technologies are important to the success of wireless communications. This book has discussed those that prove most confusing to the majority of computer users. Anyone who wishes to pursue the technologies further can obtain additional information from any of the resources listed in Appendix C.

Appendix C

Resource Guide

INDUSTRY GROUPS AND PUBLICATIONS

Andrew Seybold's Outlook on Mobile Computing P.O. Box 917, Brookdale CA, 95007, 408-338-7701

American Mobile Telecommunications Association 1835 K. Street NW, Suite 203, Washington, DC 20006, 202-231-7773

Cellular Data Packet Data (CDPD) c/o Waggener Edstrom, 6915 S.W. Macadam Avenue, Suite 300, Portland, OR 97219, 503-245-0905

Cellular Telecommunications Industry Association 1133 21st Street NW, 3rd Floor, Washington, DC 20036, 202-785-0081

Infrared Data Association (IrDA) P.O. Box 495, Brookdale, CA 95007, 408-338-0924

Mobile Data Report c/o Capital Publications, 1101 King Street, Suite 444, Alexandria, VA 22314, 703-683-4100

Personal Computer Industry Association (formerly Telocator) 1019 19th Street NW, Suite 1100, Washington, DC 20036

Personal Computer Memory Card International Assocation (PCMCIA) 103G E. Duane Avenue, Sunnyvale, CA 94086, 408-720-0107

Portable Computer and Communications Association (PCCA) P.O. Box 924, Brookdale, CA 95007, 408-338-0924

RCR 1728 Downing Street, Denver, CO 80218, 303-832-6000

Telocator *See* Personal Computer Industry Association.

VENDORS

AirSoft, Inc. 1900 Embarcadero Road, Suite 204, Palo Alto, CA 94303, 415-354-8120

Apple Computer 20525 Mariani Avenue, Cupertino, CA 95014, 408-996-1010

ARDIS 300 Knightbridge Parkway, Lincolnshire, IL 60069, 708-913-4233

AT&T Easylink 400 Interpace Parkway, Parsippany, NJ 07054, 201-331-4000

Casio Computer Company *See* Tandy Corporation

Cincinnati Microwave, Inc. One Microwave Plaza, Cincinnati, Ohio 45249-9502, 513-489-5400

CompuServe, Inc. 5000 Arlington Center Boulevard, Columbus, OH 43220, 614-457-8600

Digital Equipment Corporation 550 King Street, Littleton, MA 01460, 508-486-5503

EMBARC Communications Services 1500 Northwest 22nd Street, Boynton Beach, FL 33426, 407-364-2519 or 800-EMBARC4

Ericsson/GE Mobile Communications, Inc. 45 C Commerce Way, Totowa, NJ 07512, 201-890-3600

Ex Machina, Inc. 45 East 89th Street, Suite #39-A, New York, NY 10128-1232, 718-965-0309

Fujitsu Personal Systems 5200 Patrick Henry Drive, Santa Clara, CA 95054, 800-831-3183

General Magic 2465 Latham Street, Mountain View, CA 94040, 415-965-0400

GRiD Systems owned by AST Research, Inc., 16215 Alton Parkway, P.O. Box 57005, Irvine, CA 92619-7005, 714-727-4141

Hewlett-Packard 1000 N.E. Circle Boulevard, Corvallis, OR, 97330, 800-443-1254

IBM Personal Computer Company P.O. Box 100, Route 100, Somers, NY 10589, 914-766-3000

Intel Corporation 2200 Mission College Boulevard, P.O. Box 58119, Santa Clara, CA 95052-8119, 408-765-8080

IntelliLink 98 Spit Brook Road, Suite 12, Nashua, NY 03062, 603-888-0666

Lotus Development Corporation 55 Cambridge Parkway, Cambridge, MA 02142, 617-577-8500

McCaw Cellular Communications 5400 Carrillon Point, Kirkland, WA 98033, 206-827-4500

MCI Communications 1650 Tyson Boulevard, McLean, VA 22102, 703-415-6000

Metricom 980 University Avenue, Los Gatos, CA 95030, 408-399-8200

Microcom, Inc. 500 River Ridge Drive, Norwood, MA 02062, 617-551-1955

Microsoft Corporation One Microsoft Way, Redmond, WA 98052-6399, 206-882-8080

MobileComm 1800 East County Line Road, Ridgeland, MS 39157, 601-977-0888 or 800-685-5555

Motorola Inc. Altair Products, 50 E. Commerce Drive, Schaumburg, IL 60173, 800-233-0877

Motorola, Inc. Subscriber Products Division, Wireless Data Group (InfoTac), 1201 E. Wiley Road, Suite 103, Schaumburg, IL 60196, 708-576-1600

Motorola Inc. Messaging, Information and Media Sector, 1301 E. Algonquin Road, Schaumburg, IL 60196, 708-576-5000

Mtel's Nationwide Wireless Network (NWN) 200 S. Lamar Street, Jackson, MS 39201, 601-944-7209

National Semiconductor Corporation 2900 Semiconductor Drive, Santa Clara, CA 95052-8090, 408-721-5000

NCR Corporation 1700 S. Patterson Boulevard, Dayton, OH 45479, 800-CALL-NCR

NEC Technologies, Inc. 1414 Massachusetts Avenue, Boxborough, MA 01719-2298, 508-264-8000

NexTel (formerly Fleet Call, Inc.) 201 Route 17 North, Rutherford, NY 07070, 201-438-1400

Nomadic Systems Inc. (SmartSync) 300 Ferguson Drive, Suite 200, Mountain View, CA 94043, 415-335-4310

Novell, Inc. 122 E. 1700 South, Provo, UT 84606-6194, 801-429-7000

O'Neill Communications, Inc. Momouth Junction, NJ, 908-329-4100

Philips Electronics North America 100 East 42nd Street, New York, NY 10017-5699, 212-850-5000

Photonics Corporation 2940 N. First Street, San Jose, CA 95134, 408-955-7930,

Proxim Inc. 295 North Bernardo Avenue, Mountain View, CA 94043, 415-960-1630

Qualcomm, Inc. 1055 Sorrento Valley Road, San Diego, CA 92121, 619-587-1121

Racotek, Inc. 7401 Metro Boulevard, Suite 500, Minneapolis, MN 55439, 612-832-9800

Radio, Computer and Telephone Corporation (RC&T) 13911 Ridgedale Drive, Suite 403, Minnetonka, MN 55305, 612-542-1081

RadioMail Corporation P.O. Box 1206, Menlo Park, CA 94026-1206, 800-597

RAM Mobile Data 10 Woodbridge Center Drive, Woodbridge, NJ 07095, 908-602-5500

SkyTel 1350 I Street, NW, #1100, Washington, DC 20005, 202-236-5277

Sony Electronics, Inc. 1 Sony Drive, Park Ridge, NJ 07656, 201-930-7834.

Symbol Technologies 116 Wilbur Place, Bohemia, NY 11716, 516-563-2400

Traveling Software Corporation 18702 N. Creek Parkway, Bothell, WA 98011, 206-483-8088

Wireless Access 125 Nicholson Lane, San Jose, CA 95134, 408-383-1900

Glossary

800-MHz Band, 800–940 MHz: The section of the radio spectrum in which portions are used for cellular phones, Specialized Mobile Radio (SMR) services, the RAM Mobile Data and ARDIS networks, nationwide paging, unlicensed local area networks, and similar applications.

Acknowledgment: Acknowledgments are responses to individual packets of data or an entire message. They indicate to the sending system that the information has been received error-free.

Advance Mobile Phone Systems (AMPS): AMPS is the system used in U.S. cellular phone systems for analog cellular phone service. It is the lowest common denominator for U.S. cellular systems.

Alphanumeric: The combination of alphabetic characters and numbers. When used in conjunction with wireless communications, one needs to know if a receiver is capable of receiving and displaying alphanumeric information or only numbers (numeric).

America Online: An electronic information service available via dial-up modem. Provides information, bulletin boards, and news for a monthly fee plus hourly access charges.

Amplitude Modulation (AM): A type of modulation in which the carrier remains constant and audio is superimposed within the limits of the carrier. Amplitude modulation is often simplistically described as varying the amplitude (size) of the carrier from a zero power level to a peak power level.

Analog Cellular: Circuit-switched voice telephone communications via cellular radio channels. Analog Cellular is the current cellular telephone technology used in the U.S. It is referred to as AMPS (*see* AMPS).

Analog System: A system that uses analog signals as opposed to digital signals.

Antenna: An antenna is used to receive radio signals. In its most basic form, it consists of a simple wire that has a length relationship to the signal it is designed to receive.

AppleLink: An electronic mail and information service operated by Apple Computer Corporation. Started as an internal e-mail system, AppleLink was expanded to include all Apple dealers and software and hardware vendors. Its scope has been expanded to include the general public for a sign-up fee and hourly connect charges. AppleLink is accessible by dial-up modem connection.

ARDIS: A wireless two-way data network jointly owned and operated by Motorola and IBM. ARDIS was first installed to provide IBM field representatives direct access to their databases and messaging systems. ARDIS is currently the largest provider of wireless two-way data connections with approximately 40,000 users nationwide.

ASCII Text: American Standard Code for Information Interchange. ASCII is a uniform code used by computer and data communications systems for translating bits into bytes. The most common 128 characters are represented by a 7-bit code; an 8th bit is used for graphic and other seldom-used characters.

Backbone: The core or infrastructure of a network. The portion of the network that transports information from one central location to another central location where it is off-loaded onto a local system. In telephone terms, a local Bell operating company would off-load a long distance call to AT&T's backbone. The backbone would route the call to the intended party.

Band: A group of radio channels that is classified by the FCC for a specific purpose. Examples are the cellular band and the 10-meter amateur radio band.

Bandwidth: The range of a channel's limits. Standard two-way radio operates at a bandwidth of +/- 5 KHz, which means that the audio or data transmitted must stay within a 10-KHz limit around the specific frequency (from 5 KHz below to 5 KHz above). This is a prime limiting factor for data transmission speeds in wireless transmissions. The wider (broader) the bandwidth, the faster data can be sent.

Base Station/Fixed Point Services: Radio transmitters and receivers that are stationary. Two-way radio systems are usually licensed as "base and 10" or "base and 20," which means that the system consists of a base station and 10 or 20 mobile units.

Baud Rate: The rate at which the modem is changing the state of a line of data. Baud and bits per second are often used interchangeably to indicate the speed of data sent via modem. 9600 baud and 9600 bps are not the same measure of speed, but they have become synonymous in the computer world.

Big LEO: Low Earth Orbiting Satellites (LEOs) will orbit at different distances from the earth. Big LEOs' orbits are higher than LEOs' or Little LEOs' orbits. These satellites are used to relay wireless transmissions.

Bits Per Second: A bit is a single on-off pulse of data. Seven to eight bits create a single alphanumeric character, or byte. A modem operating at 2400 bps has a "raw" throughput of approximately 240 characters per second.

Broadcast: As used in wireless communications, to broadcast means to send a message to more than one unit at one time. A "broadcast" page message is sent to a group of pagers. Most pagers have a group address that is used for broadcast pages as well as an individual address.

Campus: An office complex consisting of a number of buildings; a small, self-contained group of buildings; a college campus.

Cell Switching: When moving from one location to another while using a cellular phone, the unit may leave the coverage area of one cell site and enter that of another cell site. The cellular system is designed to switch the call to the new cell with no noticeable drop in the conversation. Cell switching is sometimes called "handing off." While not noticeable in voice communications, the approximate 300 milliseconds this switching takes can prove to be a problem in data transmission.

Cells: A group of radio transmitters and receivers designed to provide wireless coverage to a given area. The phrase "cellular phone system" is derived from its use of cells.

Cellular Digital Packet Data (CDPD): A method of transmitting data on unoccupied cellular voice channels. The CDPD consortium includes IBM, McCaw Cellular and many "Baby Bells." Several trial systems are being installed and many should be functional by the end of 1994.

Cellular: Usually refers to cellular telephone systems installed around the world. The term is derived from the concept of a type of service that makes use of a number of "cells" in each area to provide reliable radio telephone service and frequency re-use.

Central Processing Unit (CPU): A semiconductor chip that is the heart of a computer system. Intel's 8086, 80386, and 80486 chips are CPUs.

Channel: A communications link. A channel can be a single frequency or a pair of frequencies. Cellular channels use pairs of frequencies, one for transmitting, and one for receiving.

Circuit-Switched System: Establishes a dedicated transmission path from the initiation to the completion of a call.

Client: Half of a "client-server"-oriented computer network system. Programs and information reside on the server. The client connects to the server for network access.

Code Division Multiple Access (CDMA): CDMA is a proposed method of coding that will provide digital voice (and data) by use of a wide frequency channel, on which multiple messages can be sent. Messages are decoded at the receiving end. Channel capacity is claimed to be in the order of ten times that of analog cellular systems.

Coded Squelch System: In this system, a sub-audible tone in the range of 67 to 203.5 Hertz is transmitted with two-way radio audio. A tone opens receivers that have the same tone designation. Using this system, users of multiple radio networks can operate on the same channel without hearing conversations not intended for them. The coded squelch system was developed by Motorola. Other terms include CTCSS, PL or Private Line (Motorola), CG or Channel Guard (General Electric), and QC or Quiet Channel (RCA). When radio bands became even more crowded, Motorola pioneered a new tone squelch system called DPL, or Digital Private Line, which makes use of many more discrete digital tone sequences for the same purpose.

Command Set: A set of commends used for controlling an external device. For example, "modem command set" or "AT command set."

Compression: Shrinking the size of the data to be stored or transmitted. If data is compressed by a 3 to 1 ratio, the size of the file is only 1/3 its original size. Compression techniques usually require a program at the receiving end to decompress the data.

CompuServe: One of the oldest and best established dial-up information services. Access is provided to news, travel information, and Wall Street. Most of the computer industry Special Interest Groups (SIGs) are found on CompuServe.

Continuous Tone Coded Squelch System (CTCSS): *See* Coded Squelch System.

Crystal: A device that uses quartz crystals (or synthetic quartz) to establish absolute frequency in a computer or a radio. Voltage is applied to the crystal causing it to oscillate at a specific rate, generating a highly stable source of frequency.

Data Pipe: Any means over which data is transmited. A data pipe can be a wired connection or a wireless channel.

Data Rate: The rate at which data is sent and received. *See* Baud and Bits Per Second.

Database: A file or collection of information which has a common format and can be accessed in random order by the database program that created it.

Dedicated Data Channel: A channel specifically reserved for data-only transmissions; data and voice are not mixed. The RAM Mobile Data and ARDIS wireless networks provide dedicated data channels.

Dedicated System: A system that is specifically limited to the type of service it provides. *See* Dedicated Data Channel.

Digital Data: Data that can be measured in discrete, exact values versus analog data which is information represented along a continuous range where there are an infinite number of possible values.

Digital Private Line: *See* Coded Squelch System.

Digital System: A system such as a computer or wireless network that is specifically designed to work with digital data.

Directory: In the computer industry, a directory is a listing of files contained on a disk that is read by a computer. When used in the context of electronic mail, directory refers to a listing of e-mail addresses.

Disk Operating System (DOS): The main program used by a computer. DOS is loaded when the computer is powered up. It permits applications to run on a computer. MS-DOS is the DOS developed by Microsoft.

Duplex: Duplex transmissions make use of two different frequencies, one for transmitting and one for receiving. A duplex circuit or channel is one over which both sides of a conversation can be heard at the same time. A standard telephone demonstrates a duplex connection.

Electromagnetic Radiation: An electromagnetic wave travels (radiates) through atmosphere and free space at a speed of 186,000 miles per second. The wave is made up of electric and magnetic fields and propagates by regular variations in these fields.

Electromagnetic Spectrum: The continuum of electromagnetic signals from the longest wavelengths (lowest frequencies) to the shortest wavelengths (highest frequencies).

Electronic Ink: Pen-input computing devices permit users to make notations and draw on the screen of a computer. Markings on the screen that are stored or used just as they were created are called "electronic ink."

Electronic Mail (E-mail): Information that is sent from one computer user to another over a network. Many computer systems that are tied together over a LAN provide e-mail between users of that LAN. Public e-mail services enable electronic interchange with people from different companies through wired or wireless access.

Electronic Serial Number (ESN): A code specific to a given cellular phone that is stored in ROM (Read-Only Memory) so that it cannot be changed. Along with the phone number of the phone, the ESN is used to verify that a specific unit is authorized to use the cellular system it has accessed.

Error Correction: A method whereby information sent from one system to another can be verified and corrected after it is received. Two basic methods of error correction are in use today. One requires the sending station to resend data when errors are detected. In the other scheme, the string of data includes enough additional information that errors can be discovered and corrected at the receiving end.

Ethernet: A LAN topology and wiring scheme that permits computers to be connected with wires to communicate to and from a server.

Facsimile: A standard method of transmitting graphics over a telephone line. A document is scanned at the transmitting end, sent over the telephone line as data bits, and reconstructed at the receiving end. Group 3 is the current standard; it reflects an improved level of speed and quality of transmission.

Fax: *See* Facsimile.

Federal Communications Commission (FCC): The FCC is the federal agency charged with allocating frequencies and establishing rules and regulations for use of spectrum allocated for use by the private sector. Such uses include commercial broadcast, cable television, paging, two-way radio, and many more. Spectrum used by the federal government is administered by the NTIA.

FLASH: Computer memory that combines attributes of both Random Access Memory (RAM) which can be written to numerous times but loses the data when power is turned off, and Read-Only Memory (ROM) which once written to retains the data forever. FLASH can be written to many times and it retains the data when power is turned off.

Frequency Hopping: A technology employed in Cellular Digital Packet Data (CDPD) systems which causes data signals to "hop" from one unused voice channel to another. Frequency hopping is a key element in the efficiency of CDPD.

Frequency Modulation (FM): A method whereby the radio frequency is varied when audio is applied. Two-way radios use frequency modulation. FM is also used by broadcast stations in the FM broadcast band. An FM station has a center (assigned) frequency, 107.5 MHz for example, and the audio varies this frequency plus or minus the assigned channel by a specific amount depending upon the audio.

Frequency: The number of oscillations, cycles, or events per unit of time. Frequency is measured in Hertz, which is defined as cycles per second. Frequency also refers to the portion of the spectrum assigned to a specific channel. Tuning to a specific frequency permits reception of a specific transmission.

Gateway: A link between two systems. For example, when an e-mail message is sent over RadioMail to a someone on MCI Mail, the message is routed from RadioMail to the Internet via one gateway, and from the Internet to MCI Mail via a second gateway.

Gigahertz (GHz): A measure of frequency that is equal to 1,000 Megahertz or 1,000,000 Hertz. The cellular telephone band is just below 1 GHz at 900 MHz, and PCS systems will operate in the 1.8 to 2.2 GHz band.

Graphical User Interface (GUI): A computer interface that permits users to work with graphical images rather than with text commands. GUI is pronounced "gooey." Microsoft Windows and Apple's System 7 are the most popular examples of GUIs.

Handheld Computer: A small, lightweight, battery-operated computer that can be held in one hand. Examples of handheld computers are Personal Digital Assistants (PDAs), HP 95 and 100LXs, Tandy and AST Zoomers, and Poqet PCs.

Hardened Site: A building that has been constructed to withstand the harshest of nature's storms as well as acts of war or sabotage, that houses equipment that has been designed to be fault-tolerant. Hardened sites have thick concrete walls, no windows, and several levels of redundant power (commercial, battery, and generator). Multiple routes into and out of the buildings are usually provided for telephone wires and radio systems.

Harmonics: The dictionary definition of a harmonic is "an oscillation whose frequency is an integral multiple of the fundamental frequency." Harmonics are usually unwanted by-products generated by a transmitter, and regulations require that they be minimized. The harmonic of a computer's crystal-controlled clock could be high enough in frequency to interfere with a radio receiver located in close proximity.

High Band: The designation given by radio technicians to the frequency band of 150–172 MHz. The high band is licensed to paging and radio systems, police, fire, local government, business, maritime, and commercial users.

High Power: Usually refers to the output of a transmitter. For comparison, cellular systems use very low power (600 milliwatts) for short range coverage, while paging transmitters are high power (300 watts) because of the wide areas they must cover.

Horizontal Market: The segment of a market that is generic in nature. Products designed for a horizontal market can be used for general applications that are common to most businesses, as opposed to products that perform specific tasks required by vertical or specialized markets.

Impedance: The presence of resistance in a given circuit. For antennas to function properly, their impedance must be matched to that of the transmitter/receiver. Impedance mismatch between circuits causes some of the energy to be dissipated as heat rather than being transferred.

In-bound Channel: The channel that carries data from the user to the system. The opposite of out-bound channel, which is the channel used to send information from the system to the user.

Industrial, Scientific, and Medical (ISM) Bands: Bands available for unlicensed low-powered transmitters used for in-building LANs as well as scientific monitoring devices.

Information Systems (IS): The current term for what used to known as MIS or Management Information Services, a department found within every large company. The IS department is responsible for all computing functions within a company and usually "owns" the mainframes, minicomputers, and networks to which users are connected.

Infrared (IR): Electromagnetic radiation (like radio waves) that is made up of wavelengths shorter than visible light and longer than microwave signals. IR is not visible to the human eye. Data is transmitted via IR by modulating the IR signal. A typical example of an IR device is a television remote controller.

Infrastructure: The equipment behind the network. A telephone company's infrastructure is a series of wires and switches used to provide telephone line communications.

Input Device: A device used to provide data or commands to a computer. Keyboards, pens, mice, and voice are input devices that are used to interface between a user and a computer.

Intelligent Network: A network controlled by computer switches. Routing across and within an intelligent network is handled by the network itself. The telephone network is an intelligent network, even though the user must dial a number to place a call. The RAM Mobile Data and ARDIS networks are intelligent networks that provide automatic user connections based on the name of the user and the type of service to which he or she subscribes. The user does not have to "dial" a number or request a connection.

Interference: An unwanted signal, anything that interferes with the primary communications link. Interference can be noise, another signal, or a combination of externally generated signals that occur on the channel in use. Interference problems will sometimes slow a network's operation since packets must be sent repeatedly until they are received error free.

Internet: The best definition for the Internet is from the book *Connecting to the Internet* by Susan Estrada: "The Internet is not one place or one company. It is really a descriptive term for a web of thousands of interconnected networks. You wouldn't be far off if you imagined the Internet as a kind of computer amoebae, reaching out and connecting separate islands of computer resources into a seamless web."

Iridium: Conceived and designed by Motorola, Iridium is a series of satellites that will orbit the earth (LEOs) and provide voice, data, and video communications on a global basis. The name was chosen because Iridium is the 77th element. The system was originally to be comprised of 77 satellites.

K Band: Electromagnetic frequencies in the range of 10.9 GHz to 36 GHz, mostly used for radar and satellite systems.

Kilo (K): Kilo is the designator for 1000.

Kilobytes Per Second (KBps): As a data speed measurement, KBps represents one thousand bits per second.

Licensed: A radio frequency on which a user or network provider is required to hold an operator's licensed issued by the FCC.

Local Area Network (LAN): A group of personal computers that are interconnected to share resources such as peripheral devices, files, data, and programs. There are various types of LANs, some requiring sophisticated wiring and software control.

Low Band, 30–50 MHz: The initial set of two-way radio frequencies used for police, fire, and business communications.

Low Earth Orbiting Satellite (LEO): Some new voice and data systems such as Iridium will make use of satellites placed in low orbits to provide coverage anywhere in the world.

Mailbox: A specific address for a user on an electronic mail system.

Mbps: Megabits (millions) of bits per second.

Mega: The term for 1 million

Megahertz (MHz): 1 Million Hertz. 1 MHz equals 1,000,000 Hz or 1,000 KHz.

Memory: Temporary storage within a computer. Memory chips can store information in electronic form.

Mesh Network: A wireless network that includes a number of "smart" nodes that can determine the other nodes with which they can communicate. A mesh network contains its own intelligence at the node level and does not need to be attached to a master computer; a cellular network must be controlled by a master computer.

Metropolitan Area Network (MAN): Similar to a LAN except that a MAN provides interconnection anywhere within a given metropolitan area, usually via wireless communications.

Metropolitan Statistical Area (MSA): An urban area as defined by *Rand McNally*. The FCC uses MSAs as a guide to determine coverage areas for cellular systems and soon for PCS systems.

Microwave: Electromagnetic frequencies in the range of 1 GHz to 1 THz. Microwave frequencies have historically been used for point-to-point communications, but some (1.8 to 2.4 MHz) will be used for Personal Communications Services.

Milliseconds (Ms): One-thousandth of a second.

Milliwatts (Mw): One-thousandth of a watt of power. Cellular phones have an output power of 3 watts (3,000 Milliwatts), while handheld phones operate at 600 Mw or 0.6 of a watt.

Mismatch: A mismatch occurs when a radio circuit and an antenna have not been designed to work together. The results of a mismatch are that some of the radio energy is dissipated as heat rather than as radio waves.

Mobile Data: One-way or two-way information that is sent to or received by units that are outside an office environment.

Modem: Derived from the words modulator and demodulator, a modem is a device that converts binary code used in a computer into analog sounds used by a standard telephone system and vice versa. A modem is necessary to send data over any analog circuit, whether wired or wireless.

Modulation: The process of varying some of the characteristics of a carrier wave in accordance with a signal to be transmitted. A carrier wave is a wave having at least one characteristic that can be varied from a know reference. AM radio transmitters employ amplitude modulation, while FM radio stations employ frequency modulation. Most analog radio services (including cellular phones) use a form of FM or phase modulation.

Motherboard: The main circuit board in a computer or other electronic device. Boards that can be plugged into the motherboard are referred to as daughter boards or daughter cards.

Multiplexing: A method of transmitting two or more narrow band channels over a broadband channel. The two most popular methods to accomplish this are frequency-division multiplexing, in which the board channel is sub-divided into multiple narrow channels, and time-division multiplexing where the data to be transmitted is sent sequentially, in slices of time, on the same channel.

Network: Any wired or wireless system that provides connections to multiple points and devices. The telephone system network is a system of wires, fiber optic cables, and microwave links that permit the connection of one phone to another.

Node Controller: The device within the network that is responsible for ensuring that a node has access to the network. The node controller routes information to a node.

Node: Any device connected to a network.

Noise: Any electrical disturbance that interferes with the reception and reproduction of electric signals and causes deterioration of signal fidelity. Noise can be caused by many factors including the device itself, other signals on the same channel, or strong signals on nearby channels that bleed over. Noise on circuits will cause slower information flow in proporation to the level of the noise.

Notebook Computer: A sub-segment of portable computers. The term notebook computer has come to represent computers that weigh between 5 and 7 pounds and are battery-operated.

Object Oriented: *See* Graphical User Interface (GUI).

Ohms: A measure of the resistance of a given device. Everything a signal travels through has a certain resistance, limiting the current or other characteristic of the signal.

One-way: Used to describe radio systems where information is transmitted from a central station outward to users, but users cannot transmit back to the system. AM, FM, and TV stations use one-way transmissions, as do paging systems.

Open Systems Interconnect (OSI): A seven-layer standard created by the International Standards Organization that is designed to link all types of devices from all types of manufacturers.

Optical Character Recognition (OCR): The ability of a device to decipher graphical images and convert the image (such as a fax) to text characters.

Oscillator: In its most basic form, an oscillator converts DC power to AC power. AC oscillations are determined by the time constant of the circuit. Oscillators are used to generate frequencies in radios and to provide a stable reference point within a computer.

Out-bound Channel: The channel used to transmit from a main source to a mobile unit.

Packet-switched Systems: Data transmission systems in which information is divided into packets. Many packets from many users can share the same circuit. LANs use packetized transmission methods as do most wireless data networks.

Packet: A message is divided into pieces. Each piece (packet) contains data, a destination address, sequencing codes to put the pieces back together, and error correction codes.

Paging: The use of an outbound channel over which specific tones are sent to turn on a specific receiver. Information is sent to the pager once the receiver has been alerted. Information can be in the form of voice, tones, numeric, and alphanumeric characters.

Palmtop: A palmtop is a computer is a computer that is designed to be held in the palm of the hand. Palmtop systems do not have large screens; if they have a keyboard, it is much smaller than a standard computer keyboard.

PC Cards: PC cards are used in portable and mobile computers to provide extra memory, modem and fax modem, paging, and other functions. In form, a PC card looks like a thick credit card. If a PC card meets the specifications of the PCMCIA, it should be usable in any device that is also PCMCIA-compliant. PC cards are also called IC cards (Integrated Circuit cards).

PC: Generic abbreviation for a Personal Computer.

PCMCIA Type I: The original PCMCIA specification which was limited to RAM and ROM card characteristics.

PCMCIA Type II: The current standard for the 5-mm-thick cards. Type II includes specifications to empower modems, fax transmission and reception, communications connections, and Execute In Place (XIP) cards.

PCMCIA Type III: This specification is the same as the Type II specification, except that it allows for cards to be twice as thick (10 mm), providing additional room inside the card for hard disk drives and other devices that cannot be built into the thinner cards.

Peer-to-Peer: Systems that are designed to provide communications directly from one node to another without using a network server.

Personal Communications Network (PCN): *See* Personal Communications System.

Personal Communications System (PCS): A new service that will occupy the 1.8 to 2.4 GHz frequency range (in the U.S.). PCS will provide voice and data access for end users no matter where they are. Further into the future, video access will also be available.

Personal Communicators: Another name for PDAs, but indicating more emphasis on the communications aspects of the device.

Personal Computer Memory Card International Association (PCMCIA): The PCMCIA is responsible for the specifications and standards for PC cards (IC cards) and PC card sockets that are built into computers.

Personal Digital Assistants (PDAs): This term was given to handheld computing devices by John Sculley of Apple Computer when he announced the Newton concept. PDA is one of many terms used to describe such devices.

Personal Information Communicator (PIC): *See* Personal Communicator.

Personal Information Management (PIM): A combination of applications that enables a user to coordinate his or her calendar, phone lists, to-do list, and in some cases, write short memos.

Personal Information Terminals (PITs): A term for Personal Digital Assistants.

Point-to-Point: Communications circuits designed to provide service from one specific point to another. Microwave systems operate as point-to-point systems and use antennas optimized for directional transmissions. "Fixed-point" services refer to point-to-point microwave communications.

Polarization: Refers to the type and position of an antenna. Horizontal, vertical, and circular polarizations refer to the position of an antenna in relationship to the ground. Systems work best when the polarization of the transmitter antennas and the receiver antennas are the same.

Portable Computer and Communications Association (PCCA): A non-profit organization of member companies from both industries that is working toward providing wireless industry standards.

Power Supply: A device that receives standard AC line power (or battery power) and generates multiple voltages for driving electronic equipment. Most PCs have a power supply that converts 110 VAC to 5 Volts +/- DC and 12 volts +/- DC.

Printed Circuit Board (PC Board): A laminated board of synthetic material on which electronic components are mounted. The components are connected by traces of silver on the board that replace wire.

Private Line (PL): *See* Coded Squelch System.

Prodigy: A dial-up, online information service like CompuServe. Prodigy was designed primarily for individual users and is a joint venture between Sears and IBM.

Proprietary: Refers to a standard or technology that is protected by law (patented), that is used by a specific company. Proprietary information or techniques can often be licensed.

Protocols: Standards for procedures. Voice and data protocols specify standards that enable software and hardware vendors to design and make use of information from outside sources.

Public Switched Telephone Network (PSTN): This term describes the present wired telephone network.

Push-To-Talk (PTT): A button or switch located on a microphone or console that turns on the transmitter when the button is depressed.

Radio Conference (WARC): *See* World Administrative Radio Conference.

Radio Receiver: A device designed to convert information sent through the ether into intelligible information. The simplest of these is an AM broadcast band receiver.

Radio Site: Where radio equipment is located. Two-way radio systems use high towers or mountain tops to be able to transmit over wide distances. Cellular systems use a number of sites that are located closer to the average terrain to limit the distance covered by each cell.

Radio Spectrum: The portion of the electromagnetic spectrum over which radio waves can be transmitted and received. A sub-set of the electromagnetic spectrum.

Radio Transmitter: A device designed to send (transmit) information through the atmosphere. A typical transmitter generates a specific frequency and then imposes information on the channel. When the information is received, is converted back to its original form.

RAM Mobile Data: One of two companies that provide nationwide data-only wireless networks within the U.S. (*see* ARDIS). RAM also operates systems in Europe and Asia.

Random Access Memory (RAM): Memory contained in a computer which retains information only as long as power is applied. Information can be written to RAM many times.

Read and Write Memory: *See* Random Access Memory.

Read Only Memory (ROM): Memory in an integrated circuit which contains information or programs that cannot be erased. ROM retains its memory when power is turned off.

Receiver Sensitivity: The ability of a receiver to "hear" radio signals. More sensitivity in a receiver results in better performance.

Regional Bell Operating Companies (RBOCs): Companies that AT&T was forced to turn loose after the break-up of AT&T by the courts. RBOCs are limited at present and can only provide local telephone access. They must hand off long distance calls to one of the three long-distance carriers (AT&T, MCI, Sprint).

Repeater: A radio transmitter and receiver configured in such a way that the signal received by the system is immediately re-broadcast by the transmitter (repeated). Repeaters are used to increase the range of mobile units in a fleet.

RJ-11: The small telephone connector used on most wired telephones. An RJ-11 connector supports up to 4 wires; only two are normally used for telephone systems.

Roaming: The operation of a wireless unit (cellular phone, data device) outside of a customer's prime area of operation. In cellular systems, a "roamer" is operating out of his or her prime coverage area and must pay special roaming charges.

Seamless: A type of roaming where the user does not have to take any specific action to use a wireless device outside of its prime area. Seamless roaming implies that users will be able to operate the device in the same manner no matter where they are, and that they will receive a single bill for all charges.

Serial Cable: A cable designed to interface the serial port of a computer to an external device such as a modem, wireless modem, or printer. There are two standard connectors for serial cables: 25 pins and 9 pins.

Server: A central computer or device in which the master network software resides. The server controls the network and all nodes must be connected to the server to use the network.

Service Area: The area the FCC has licensed to a carrier to provide wireless service. The FCC divides the country into Metropolitan and Rural Services areas (MSAs and RSAs) for the purposes of cellular systems.

Session-based: A wired or wireless connection that is made between a user and a specific location. The connection must remain constant during the entire exchange of information. Telephone conversations are "session-based" transactions. While a session (conversation) is taking place, no one else has access to either party participating in a session connection.

Side-band: A frequency band on the upper and lower side of the carrier frequency that contains the frequencies generated by the process of modulation. In some cases, sidebands are not wanted and must be eliminated. In others, sidebands can carry important additional information. For example, FM stereo systems send the stereo portion of their broadcasts on a sideband of their assigned channel.

Simplex: In simplex communications, both parties use the same channel. While one person is transmitting, the other is listening. The transmitting party cannot hear other users on the system until he or she stops transmitting.

Simulcast: In radio communications, simulcast refers to a method where two or more transmitters send the same information at the same time. Special engineering techniques must be used to make simulcast systems work properly.

Site: Refers to a specific location where radio equipment is located.

Sleep Mode: A low-power mode designed into computers. In sleep mode, minimal current is required to keep the system on. When a key is depressed, the system "wakes up" and is instantly ready for use. Sleep mode is not to be confused with a computer screen-saver application which merely replaces the screen image to prevent damage from image burn-in.

Specialized Mobile Radio (SMR): A system that uses a series of channels that are controlled by a central computer to permit more mobile units per channel than can be handled with conventional radio systems. First licensed by the FCC in the early 1980s, SMR systems are also called "Trunked Radio Systems." The majority of SMR systems are used for dispatching fleets of vehicles, but some also provide mobile telephone access. Many SMR operators are implementing or planning expansion of their systems to compete directly with cellular and data-only networks such as those of RAM Mobile Data and ARDIS.

Spectrum: Electromagnetic spectrum includes colors and other waves both seen and unseen. Spectrum is measured by the number of sine waves crossing a given point in a second, with one Hertz being one sine wave per second. Radio waves are part of the electromagnetic spectrum and are regulated in the U.S. by the FCC and NTIA.

Spread Spectrum: A radio technology that makes use of wide radio channels. There are several types of spread spectrum systems in use. The technologies were first used by the government to provide secure communications links. The commercialization of these systems is designed to increase the speed of data transmission and to provide systems that are less susceptible to noise interference.

Spurious Emissions: Unwanted components of a radio transmission. Spurious emissions are limited by the FCC and all radio equipment must meet FCC requirements for the reduction of spurious emissions.

Star Network: A wired or wireless network that is configured in such a way that each node communicates only through the main controller (server). Two nodes can talk to each other only by sending and receiving their signals through the central controller.

Static Read and Write Memory (SRAM): IC devices similar to RAM, except that SRAM can retain the data they contain even with no power.

Store-and-Forward: A messaging system that provides the capability to send a message to another user when the intended recipient is not online. The message is held (stored) by the system until the intended recipient comes online. At that time, the message is sent by the system (forwarded).

Sub-notebook Computer: A portable computer classification below notebook computers. Sub-notebook computers are smaller and lighter than standard notebook computers, have smaller screens, may not have full-size keyboards, and weigh less than 5 pounds.

T1 Lines (Circuit): A high speed digital data circuit that permits data to be sent and received at 1.544 Mbps.

Tablet Computer: A computer designed to make use of a pen input device in place of a keyboard. Tablet computers are about the size of a tablet of paper. Some newer tablet computers can be attached to a keyboard to become a notebook computer as well.

Telemetry: Data sent over a wired or wireless network which indicates the status of a device. Telemetry is used to monitor such things as water level in a water tank or the temperature of a pump that is at a remote site.

Terminal: A keyboard and screen designed to be connected to a host computer. A terminal does not have a main processor, memory, or disk storage capacity. These devices are sometimes called "dumb" terminals because of their lack of computer intelligence.

Text Document: A document that is not formatted by any specific word-processing program. Since most word-processors use application-specific formatting, a text document is the lowest common denominator for exchanging text information from one computer to another.

Throughput: The actual speed of the network. To calculate throughput, system-specific headers and footers, as well as any other related network-specific bytes must be removed. The resulting speed is referred to as actual throughput.

Time Division Multiple Access (TDMA): A technology for digital cellular service that increases a system's capacity by multiplexing, or time-slicing a radio signal. TDMA improves cellular system capacity by 300 percent and improves security since it cannot be understood when received on a standard radio receiver.

Token Ring: A token ring network is configured as a ring, and a "token" is sent from the server around the ring. If a specific node has data to send, it captures the token, adds the data, and passes it on. If a node has no data to send, it merely passes the token on to the next node.

Tone-Plus-Voice: A paging technique whereby tones are sent to activate a specific pager or group of pagers. Once the pagers are activated, a voice message which can be heard on the pager's built-in speaker is sent. Tone-plus-voice paging is used in many hospitals, but its use in wide area paging systems is decreasing because a voice message takes longer to send than a data message, and because voice systems do not provide the same area of coverage as that of data-only systems.

Transceiver: A combination transmitter and receiver. Transceivers are complete two-way devices designed to provide voice and/or data communications over a wireless circuit.

Transport Layer: The bottom level of the OSI model; the level over which all information is sent.

Trunked: A trunked radio system is an expansion of a repeater radio system in which multiple repeaters, on different radio channels, are connected by a computer. All base stations, mobile units, and portable units listen to a central control channel. When a request to "talk" is received by the central computer, the entire fleet of users is switched to a vacant channel. Once the conversation is completed, all units are automatically returned to the control channel. Trunked radio technologies are the basis for Specialized Mobile Radio Systems (SMRs).

Two-way: Two-way radio systems are designed to both send and receive information. Mobile and portable units in the field can hold two-way (both direction) conversations with a dispatcher and with each other. Cellular phones employ two-way techniques.

UHF Band, 450–470 MHz: The band of radio spectrum located just below UHF-TV channels. One-way and two-way radio systems operated by business, police, fire, and local government can be found in this band. Although new mobile telephones have moved off this frequency, older units also operated in this band.

UHF-TV, 470–512 MHz: Television channels 13 through 82 occupy this range of the spectrum. In some urban areas of the U.S., unused channels are available for use by two-way radio systems on a non-interfering basis. Such mobile systems are said to operate in the UHF-T band.

Unlicensed: The FCC has set aside some frequency bands for use by unlicensed radio devices such as garage door openers, portable phones, baby monitors, and wireless local area networks. The Citizens Band (27 MHz) is now unlicensed. Other bands include 49–50 MHz (portable phones), 902–928 MHz (wireless LANs, portable phones, and other devices), as well as bands in the 1.4 and 2.5 GHz range. Use of unlicensed bands is open to anyone and there is no protection against interference generated by other users.

Vehicle Location: The ability to pinpoint the location of a vehicle within a given range. Police departments and transit companies have used vehicle location systems for many years. Coupled with Geographic Positioning System (GPS) satellites, several companies offer systems that can track and locate vehicles anywhere within the U.S.

Vertical Market: The segment of a market that requires specific equipment or software designed for a particular group or classification of users. Examples of vertical markets include sales, service, real estate, meter readers, and police and fire services.

VHF Band, 150–174 MHz: The VHF band is located above the FM broadcast band but below VHF-TV channels 7–12. After the low band was completely filled with two-way radio band allocations, the FCC opened the VHF band for one-way and two way radio use.

Watt: A measure of power of a transmitter.

Wide Area Network (WAN): WANs are similar to LANs, but WANs cover a much larger geographic area. In a WAN, a computer may be connected via dial-up, or dedicated wireline, radio, and microwave circuits.

World Administrative Radio Conference (WARC): The world body charged with frequency allocations on a world-wide basis. The FCC can only regulate frequencies allocated. .

X.25 Telephone Line: A CCITT standard governing interfaces on packet-switched networks.

X.400: A CCITT standard for the exchange of electronic messages between on-line systems.

Index